PRAISE FOR

Michael Perry and **TRUCK**: A LOVE STORY

"Like his previous book, Perry once again animates his friends and neighbors on the page, bringing them to us with keen wit and respect. . . . As comfortable at a NASCAR race as he is at New York City's Whitney Museum of American Art, Perry blends macho yin with sensual yang. . . . Beneath the flannel surface of this deer-hunting, truck-loving Badger is the soul of a poet and a man at work balancing his masculinity against his softer urges toward food, sentimentality, and literature. It is this tension, this misfitting of parts, that creates a delicious tautness in his potent memoir. . . . Perry takes each moment, peeling it, seasoning it with rich language, and then serves it to us piping hot and fresh."   —*Chicago Tribune*'s "Books"

"*Truck: A Love Story* is a touching and very funny account of a year in Mr. Perry's life in rural Wisconsin. . . . Mr. Perry's descriptions and characterizations of the inhabitants of his community make them thoroughly engaging, and his commentary on his restoration is equally informative."   —*New York Times*

"*Truck: A Love Story* is the perfect present. . . . Wonderfully evocative. . . . Buy a stack of these books for the elusive guys in your life."   —*Washington Post Book World*

"Even though *Truck: A Love Story* is one of the most wonderfully romantic love stories you're ever going to read . . . don't discount this book as a gift for a man. . . . It's also a darn fine book about trucks, gear-heads, garages, and gardening. . . . Don't miss this book. It's that simple." —*Brownsville Herald*

"Humorous without being cynical and heartfelt without being overly sentimental, Perry is my kind of memoir writer."

—ShelfAwareness.com

"Perry's narrative voice—smooth and low-key—invites readers along for what turns out to be a most pleasurable ride."   —Nancy Pearl

"Revisiting New Auburn, Wisconsin, the scene of his fine 2002 memoir *Population: 485* (now 562), Perry's follow-up is a comforting slice of literary meatloaf."   —*Outside* magazine

"Politicians would do well to study Perry's writing to gain a better understanding of the disconnect between rural and urban, left and right. The rest of us should read it because it's witty and wise."
      —*Capital Times* (Madison, Wisconsin)

"There is so much to like about this book. . . . Perry has just the right touch—part Bill Bryson, part Anne Lamott, and with a skim of Larry the Cable Guy and Walt Whitman creeping around the edges."
      —*Lincoln Journal Star*

"A 2002 review of *Population: 485* in the *Oklahoman* said, 'If you read only one nonfiction book this autumn, make it this one. It's that good.' Four years later, I'll repeat that first line but change the second to: 'This one's even better.' "   —*Oklahoman*

"In the end, the truck is restored, Perry is restored, and, to be honest, we who read the book are restored. It's another wonderful book by a great Wisconsin writer."   —*Wisconsin State Journal*

TRUCK: A LOVE STORY

TRUCK

A LOVE STORY

To Noah –
Double-Clutch!

Michael Perry

Michael Perry

HARPER ● PERENNIAL

NEW YORK ● LONDON ● TORONTO ● SYDNEY

HARPER ● PERENNIAL

A hardcover edition of this book was published in 2006 by HarperCollins Publishers.

P.S.™ is a trademark of HarperCollins Publishers.

FIRST HARPER PERENNIAL EDITION PUBLISHED 2007.

Designed by Renata Di Biase

The Library of Congress has catalogued the hardcover edition as follows:
　Perry, Michael.
　　Truck : a love story / by Michael Perry.
　　　p.　cm
　　ISBN 10: 0-06-057117-9
　　ISBN 13: 978-0-06-057117-7
　　1. IHC trucks—Maintenance and repair—Popular works. 2. Antique and classic trucks—Popular works. 3. Popular culture—United States—Popular works. 4. New Auburn (Wis.)——History—Popular works. 5. Perry, Michael, 1964——Homes and haunts—Popular works. I. Title.
　　TL230.5.I223P47　2006

　　629.223'2—dc22　　　　　　　　　　　　　　　2006043394

ISBN: 978-0-06-057118-4 (pbk.)
ISBN-10: 0-06-057118-7 (pbk.)

10　11　ID/RRD　10

To all who hold the world together
including
the shaman girl

Acknowledgments

Thanks to:

First and foremost, to my parents—anything decent is because of them, anything else is simply not their fault.

Mark, who said, "Yeah, I could help you with that." And Kathleen, obviously. Uncle Mike, you started it all. Matt Marion, for giving me steady work and two eyebrows. Tony Baker, for being right around the corner. Dan Baker for handling security and introducing me to the people of a certain island, in particular the Hennings and the Red Cup. Speaking of coffee, Racy's, always Racy's. Shimon Lindemann, my favorite possibly German hipsters. Our Colorado family. ALR, rhymes with cigar. Kris and Frank, one last look in the fridge. *Mi cunado*, for a faraway place to hide out and write. To C. Dale Young, whose writings and poetry—while not specifically cited in these pages—stood powerfully behind certain moves. "On Beethoven's Ninth" gave me two barrels of resolve.

All International people, from informal conversations to formal correspondence, and special thank-yous to the Wisconsin Historical Society, Al Pancake, Guy Fay, Lee Grady, Daniel Hartwig, and Marguerite Moran.

Men's Health, Outside, Backpacker, Hope Magazine, and *Wisconsin West,* for publishing essays from which some of the material for this book was drawn.

Alison, Jeanette, Tim, Jen, Scranton. Tina and Lisa. Blakeley, for everything that comes after the typing is done. Krister, from all over the dang map.

Frank, *without you* . . . Jayne and Chris, and miss you, Mr. B. The McDowells. Mags, whatever dimension. And everyone of the two tables: corner of the Joynt, kitchen of Taylor.

Everyone who shows up on the road, thanks.

If I missed you, say so. Gently.

And always, no matter where, "Nobbern."

AUTHOR'S NOTE

A man for whom I have great respect once told me that the power of nonfiction resides in trust—once the reader no longer trusts the writer, that power is lost. The following book is set in the years 2003–2004. Using items including calendars, notebooks, correspondence, snapshots, and even receipts from Farm & Fleet, I have done my best to assure that chronology and veracity have been honored. Mistakes may be found. In some cases, names have been changed. Because the bulk of the book was written in 2005, I sometimes reference events, conversations, and writings occurring after 2004. While all the details of my gardening are sadly true, some of those details are drawn from different years. I may also have conflated two dirt-track stock car races (I got caught up in things and my notes are unclear), but I assure you I attended both and have the hearing loss to prove it.

Regarding all references to the International Harvester Company, Irma Harding, and my truck in particular, I am not an expert, nor am I a collector. Drawing on a number of resources, I have done my best to be accurate, but I trust there will be points of disagreement among those more informed than I, as I often found disagreement in research and reference materials (for instance, on whether or not my truck was *produced* in 1951 or *sold* in 1951). I refer to my truck throughout the book as an L-120, because that is what the lettering on the side of the hood says it is, and I reference that lettering several times in the book. International made a myriad of L-models (sixty-six, if my research is accurate) and there is a plate inside the cab of my truck stating it is actually an L-122. For the purpose of sparing both the reader and writer confusion, I call it an L-120 throughout, since that's what people see on either side of the hood. I respectfully leave the finer details to the experts and hobbyists, who have been very helpful to me in this endeavor. For those of them brave enough to read this book, I suspect they will feel as if they are watching a walrus attempting to play a piano.

TRUCK: A LOVE STORY

Introduction

THE STORY BEGINS on a pile of sheep manure the size of a yurt. Dad stacked it alongside the barn one winter, and I climbed it, a fact documented in a thirty-something-year-old photograph of a miniature me waving from the rounded peak, clearly thrilled to have summited the dung. My jacket is unzipped and the sun is bright, but the landscape is blank with snow, and the shading of the sky—bleached horizon rising to a zenith of cyanotic blue, and not a cloud to be seen—suggests the day was cold and deepens the green of the tall pines in the background. My hood is up, my pants are tucked inside rubber barn boots, and I am leaning on my sawed-off pitchfork as if it were an alpenstock. I am grinning like the hick spawn of the devil and Sir Edmund Hillary.

That spring, Dad hauled the manure away and spread it on the fields. The spot where the pile once stood was marked by a circle of earth stained dark by all the good juice that had percolated down. When Mom and Dad put in their big garden across the yard that year, they gave my brother John and me some sweet corn and pumpkin seeds to plant in the stain. The plants rocketed from the earth. In another photo taken later that summer, John and I are standing barefoot before the patch. It's fairly early in the season—I can see the hay wagon docked at the elevator, meaning Dad was baling first crop—but the sweet corn has unfurled to the level of my shoulders and the pumpkin leaves are the width of a tractor seat.

I went through a long stretch of adolescence associating the garden with chores (specifically weeding), and then another long stretch of city living in which I had no garden at all, but when I moved to the village of New Auburn, Wisconsin, eight years ago and found myself in possession of a backyard, I began to get green urges. Beginning with a simple raised bed, I attempted to grow a portion of my own food. Eventually, the garden grew to include eight raised beds, a scattered collection of pots, and a mound of dirt left over when the raised beds and pots were filled.

I have proven to be not much of a gardener. For one thing, I spend too much time away from home. When it comes to gardening, there are distinct advantages to being *present*. For another thing, I lack specific knowledge and discipline. I tend to garden based on impulse and intuition. Apparently there are better ways. If my raised beds have any consistency, it is that they are anemic and squirrel-riddled. My garden gives me inner peace and salad, but it also yields cat turds and wilt. Still, my desire doesn't die. Every year, I want to plant again. And I credit the memory of that sheep manure garden. I keep believing I can duplicate it. That I can just slap some seeds in the ground and they will come busting up. If you drove down the alley behind my old house this past winter and saw the light in the basement window above my gardening bench, or saw me bent over the raised beds transplanting seedlings in the spring or yanking weeds in the summer, or if you saw steam on my windows as I blanched the skins off the last of my tomatoes, that was me living out the annual year of the garden, driven in part by appetite and thrift, but also driven by a desire to reconnect with the little boy who loved to crawl between the stalks to a quiet place where he could listen as the sweet corn leaves scraped in the breeze. To rest his chin on his hands and peek through the shadowy tangle of vines to that one spot where the sun pierced the green canopy in a bright bolt and lit a powdery pumpkin blossom up all electric orange.

That same little boy once rapped his head on the windshield of his father's old International pickup truck with such force that a sparkly spiderweb appeared in the glass. We can only guess what effect this had on the little boy, although we do know he grew up to have trouble doing fractions and once on the way home from swimming lessons at the big green lake in Chetek he grabbed the shift lever on the International and tried to downshift on the fly, which made his mother say something very sharp indeed, and to this day he is better at upshifting than downshifting, but then isn't everybody? Lately it turns out it might have been his brother who threw the shift lever and he is misremembering, to which he replies, Holy Crap, look at the size of the crack in that windshield. We do know that the little boy grew up to have an International pickup of his own, which he loves very much but has to content himself with admiring it as it sits there, because it is not running and he does not know how to fix it. All in all he is content with this state of affairs because he likes to study the old truck as the still center of all the movement around it, something along the lines of "the sleep of trees or stones," as Simone de Beauvoir put it when describing Jean-Paul Sartre's discussion of an inert chestnut tree, which we take to symbolize that the little boy is in way over his head. Returning to the adult version of me, prior to leaving on a recent extended road trip, I took my father aside and told him that were I to perish behind the wheel and the Iowa or some other state patrol were to return my belongings, he should know that while I was enjoying the twelve-cassette packet titled *No Excuses: Existentialism and the Meaning of Life*, I was not completely buying the *Meaning of Life* bit. The amateur study of philosophy is like taking a few laps with a NASCAR driver. You're not qualified to do it on your own, you have no business behind the wheel, but for a few laps or paragraphs, you're right in there with'em, and when it's all over, you've learned something. Or, as my local fire chief once said, you've simply exasperated the situation.

And it remains difficult to get a philosopher to deliver a load of pig manure to your garden. So I really should get the truck going. It sits there falling apart with a case of nuclear cradle cap, thirsty for paint and a gas tank that won't leak. The project would give me license to make numerous trips to Farm & Fleet, where the livestock section feels sadly ever more the equivalent of a hobby section, but the sign over the drinking fountain that says PLEASE NO TOBACCO JUICE remains, and consequently, so does hope. I don't expect much, and the little pleasures suffice. This morning for coffee I ground four scoops of Farmer to Farmer Guatemalan Medium and when I pulled the grinder cap and sniffed, it was all I could do not to flop right over and shake my leg like a dog.

So. The year is planned. Grow a garden and recapture my youth. That, and get my decrepit 1951 L-120 International pickup truck running in time for deer hunting season in November.

Right off the bat, I got distracted by a woman.

Chapter 1

IRMA

I HAVE THE HOTS for Irma Harding. I wish I might couch my desire in more decorous terms, but when our gazes lock, the tickles in my tummy are frankly hormonal. My feelings are beyond ridiculous and destined to remain profoundly unrequited, but I draw a wisp of comfort from the fact that I am not squandering my libidinous yearnings on some flighty young hottie. Irma Harding radiates brightness and strength. She furthermore appears to have good posture. As a younger man, I would not have looked twice at Irma Harding.

As a younger man, I was a fool.

A man learns to tune his sensibilities. Consider the eyes. Your callow swain will be galvanized by coquetry and flash; your full-grown man is taken more by the nature of the gaze. A powerful woman's eyes are charged not by color but by intent. The strong woman does not *look* at you, the strong woman *regards* you. Irma's gaze is frank, with a crinkle of humor at the crease of each eye. She knows what she is looking for, and she knows what she is looking *at*. She has a plan, and should she encounter events for which she lacks a plan, she will change gears without fuss.

In the one picture I have of her, Irma is grinning. The grin is well short of goofy, but it does pull a little more to one side than the other. Her lips are full and gracious, although some might suggest she back the lipstick down a shade. Her teeth are white and strong. The left upper incisor is the tiniest tad off plumb, but as with the faintly lopsided grin, the net effect is to make her more human, more desirable. Irma's grin is an implication, the implication being that while she would never *tell* a naughty joke, she would quite happily laugh at one.

Irma is the product of a time when a woman—even a strong woman—strove mostly and above all to please her husband. There is a danger here, a danger that you will form an image in your head of Irma as a servile drone. Look at those eyes again. They are the eyes of a woman who willingly mixes an after-work highball for hubby, but when she delivers the tumbler it is snugged in a napkin wrapped tight as a boot camp bedspread, and hubby will not underestimate the consequences pending should Irma later discover a water ring on the end table. He will droop home slack-tied and gray from the desk-job day, and she will meet him at the door crisp as a celery stick, her cheeks bright, her backbone straight. She will kiss him and take his briefcase, but he will be left to fetch his own slippers. When he settles in the big living room chair, he will turn an ear to the kitchen, from which will emanate the sounds of dinner under way. Not the clownish clatter of pans, or the careless jangle of cutlery, but the smooth *whizzz* of a blender, the staccato *snickety-crunch* of the carrot being sliced, the civilized *tunk* of the freezer door dropping shut on its seal. Lulled by these muted vibrations of efficiency, the husband will drift in the aura of provision and comfort, and his mind will ease.

But just as he is about to drowse, he hears the meat hit the pan, and he rouses to the idea that food is being cooked. He is reminded that he must daily—like any caveman—use his hands to put food in his face. He feels juices release, and his gut rumbles. And that's why Irma gets me bubbling. She may be cast as the stereotypical nuclear housewife, she may be complicit in the premise that a man is to be served, but when I lock on those eyes, I hear the sizzle in the skillet, and I know Irma knows: no matter how you tweak the parsley, eating remains a carnal activity.

○ ○ ○ ○

Two winters back, a man knocked at my front door. I like to look folks over before I step into the open, so I paused a moment to study him from behind the glass. He had backed away from the porch and was standing on the short patch of sidewalk beside the driveway. My driveway could use some work. I'm no home improvement specialist, but I admit that if you have to *mow* your asphalt driveway there's work to be done. When I opened the door, the man turned to look at me but held his place on the walk. He had one eyeball smaller than the other.

"That truck for sale?" He squeezed the small eye shut when he talked. He was pointing at the old International Harvester pickup parked in my driveway. It's been there awhile. The tires have formed depressions in the asphalt and a sapling is growing through one wheel well. The sapling is six feet tall and thick as a buggy whip.

"Sorry, nope" I said.

"That got a six-cylinder in it?"

"Yep." I hoped he wouldn't get any more specific. My capacity for mechanical minutiae doesn't go much past lug nuts. One question, and he had nearly depleted the store of my knowledge regarding the engine. Embarrassing, for a guy to have such affection for an old truck and yet know so little about it.

"I need that thing." It was a declaration, not a request. He trained his one-eyed stare directly at the truck. "This buddy of mine's got a road grader, he put a six-cylinder International engine in it. Everybody told him you can't run a grader with that little damn engine." He turned his face back to me and clamped the eye a little tighter. "Hell, he can spin every wheel on that thing." He spit, poorly. A thin string of snoose trailed in the breeze, then snagged on the stubble of his chin. It was cold enough I expected the string to stiffen and hit the ground with a faint tinkle.

"Dropped that damn grader through the ice last week. Won't run fer shit. Engine's shot. Thought maybe you'd sell yours."

I scuffed my boot and shook my head. "Nah, me and that truck go way back. Can't do it."

The truck is unpainted, punched with dents, and runs heavy to rust. I've had it for just short of twenty years. Last time I got it running, maybe six years ago, the gas tank sprung a leak and all the fuel ran out. I have an abiding affection for the truck. During several lean years, it was the only vehicle I owned. My relationship with the truck has outlasted four of my most enduring romances combined, an observation the parties now departed deserve to claim as a key element in their exoneration. I have watched that truck rust in a heap for far too long, and long to hear it run again. I want to back it out the driveway, clunk the shifter into first gear, and double-clutch my way through the gears until I'm blasting down Five Mile Road like the old days. The truck is on my to-do list. I have a busted screen door on that to-do list. It's been flapping in the breeze since the Clinton administration.

"Don't the city give you no trouble, just leavin' it set there like that?" The man was really squinting at me now.

"No . . ."

"It's a damn eyesore!"

"I like to think of it as a sort of a status symbol." I was trying for a joke. The next-door neighbors have a modest collection of decrepit beaters parked in and around their garage, and the house across the back alley could pass for a smallish salvage yard.

"Jee-zaahs Christ! I'm from Chippewa Falls! You'd never get away with that in Chippewa Falls!"

Well *la-di-da,* I thought. Too bad for you.

○ ○ ○ ○

Those of us who covet International pickups nurture a perverse sort of pride. Open the cab and you will catch a whiff of geek. As trucks go, Internationals lack the pop culture resonance of a Ford or Chevy, nor do they have the arcane appeal of the rarities—say, a Studebaker Champ. Internationals reside somewhere in the dull middle, associated more with plowed fields than the open road. The heritage of the International Harvester Company is strictly agricultural. Have you heard an International owner make affectionate reference to his truck as a *binder*? *Binder* is shorthand for *corn binder*. How hip can you be, driving a truck nicknamed after an obsolete piece of horse-drawn farm equipment? This is like nicknaming your laptop *the slide rule*. One feels the geek factor giving way to dork factor.

My old truck and Irma Harding share a common ancestry, traceable to 1831, when a group of people gathered in Lexington, Virginia, to see the inventor Cyrus McCormick demonstrate a horse-drawn reaper. In a single day, he cut six acres of grain. The numbers seem quaint now, but they caused a sensation in their time. The reaper's ability to replace gangs of men with a single machine was revolutionary—some historians rate its effect on modern agriculture as comparable to the invention of the plow. (It is a matter of grim irony that some reports indicate the horses panicked during that first field test, and the reaper was pulled instead by slaves.) The reaper was slow to catch on, but by 1870 McCormick was selling 100,000 units per year. When he died in 1884, he had accumulated a monumental fortune and left behind a roaring company destined to change the nature of agriculture forever. In 1902, stockholders merged his McCormick Harvesting Machine Company with the Deering Harvester Company to form the International Harvester Company. True to its name, the firm quickly became a worldwide powerhouse, selling like mad on the home front while exporting its products to Russia, South America, Africa, and the British Empire. By 1906, International Harvester was manufacturing anything and everything agricultural. Reapers, yes, but also plows, seeders, beet pullers, and manure spreaders. Husker-shredders and lister-planters. Pickers and planters, binders and shellers. Hay mowers and hay rakes. Balers and tedders. Grain drills and twine. Feed grinders, hemp cutters, and cream separators. The first International tractor was a natural extension of the agricultural line when it chugged out of the factory in 1906. A year later the company debuted the Auto Wagon, precursor to the modern pickup. The Auto Wagons evolved quickly, and International was soon a leader among pickup truck manufacturers. In 1951, the year my truck rolled off the lot, they were going strong, shoulder to shoulder with Ford, Chevy, and Dodge.

Leveraging its agricultural market, International had begun producing milk coolers for farmers in 1935. Ten years later, the company added a line of refrigerators for the house. Domestic freezers were added in 1948. While I grew up seeing the distinctive International Harvester logo (a red lowercase *i* superimposed on a black *H*) on trucks, tractors, and corn pickers, I was surprised the first time I saw it incorporated into the handle of a refrigerator in the back of the Spring Street Sports bike shop—in, coincidentally enough, Chippewa Falls, Wisconsin. The refrigerator is still there, covered with racing stickers and stocked with festering burrito scraps. The second was down at our local fire hall. It was painted flat green and on your honor you dropped your quarter through a slot cut in an old ammo box before taking a pop. I don't believe I've ever laid eyes on an International Harvester freezer. The chest freezer in my basement is a generic undersized cube. Made in China, bought it in a big box store. No character, but it works great. Mostly it contains vacuum-packed venison, chicken, some fish, homegrown herbs frozen in water (so that they will be fragrant when I toss them on the compost heap next summer), and occasionally slabs of smoked carp.

As was the practice of many food and appliance companies, International Harvester attempted to draw attention to its refrigerators and freezers through the publication of promotional recipe books. Just as General Mills had invented Betty Crocker, International Harvester invented Irma Harding, and put her on the cover of *Irma Harding Presents Freezer Fancies*. The cover is dominated by her black-and-white portrait. Her head is framed in a nimbus of white. Tucked tastefully to the left of her collar is a miniature copyright sign. Irma Harding is not real. I do not hold this against her. Her initials pretty much confirm what the copyright symbol implies—Irma was conceived in the advertising department and delivered by an artist on deadline. My candle burns undimmed.

Irma's career was brief. "Millions Will Follow Her Counsel and Leadership . . . Millions Will Call Her Their Friend . . . ," read the headline from the October 1948 edition of *International Harvester Dealer News,* in which Irma debuted, but in 1955, second-guessing their ability to compete in the domestic refrigeration business, company officials sold the division to Whirlpool, and Irma went the way of the horse-drawn reaper. Pickup truck sales held steady through the 1960s, but Ford and General Motors were selling in much greater and ever-increasing numbers. By the 1970s, it wasn't even close, and in 1975 International stopped making pickup trucks altogether. Tom Brownell and Patrick Ertel, authors of *International Truck Color History,* attribute the decline of International's light truck line (the company continued to make large industrial trucks) to ongoing labor conflicts, turf battles between members of management, the predominantly rural location of International dealerships, and—note for future reference—*lack of marketing attention toward women*. Noticing that women were becoming more in-

volved in vehicle-purchasing decisions, many manufacturers responded
with truck ads featuring women at the wheel—driving campers, for instance. If a woman showed up in an International ad, it was usually in
the form of an accessory.

○ ○ ○ ○

I am a sucker for the idea of a time machine, and sometimes I look at
my unmoving truck and quite unoriginally wish I could have been the
guy to whom the salesman handed the keys in 1951. I'd like to get a
look at the country back then, the big changes coming on, but still a lot
of dirt roads over which to roll. I imagine myself bumping down a set of
sun-dappled washboards, giving the truck its first coat of dust. Maybe
I'd be coming home to a wife, and maybe she'd hear me coming and be
waiting on the porch in an apron, waving a dish rag, and I would roll in
there guilt-free and anticipating beef roast and green beans, having not
yet learned I was a chauvinist piggy. Alternatively, my wife may have
been planning to gag me with the dish rag, bind me in the apron, and
take my spanking-new truck on the lam. In 1951, American women
were entering their fourth decade with the vote. They were fresh off
the Rosie the Riveter years, when millions of their sisters had shown
up for work and proven themselves in scores of traditionally male roles.
In 1951, when Earl Silas Tupper became desperate to save his failing
plasticware business, he turned not to a man but to a woman: the deliciously named Brownie Wise. The company shortly posted revenues
of $25 million, Tupperware entered the national lexicon, and Wise
became the first woman ever featured on the cover of *BusinessWeek*.
The 1951 debut of *I Love Lucy* put Lucille Ball on a trajectory that
would culminate in her reign as the first female studio head in Holly-
wood history. The first Pulitzer Prize for Foreign Correspondence ever
given to a woman was awarded in 1951, when journalist Marguerite
Higgins was recognized for her Korean War reporting—the implication being that a girl with a pen might aspire to more than taking dicta-
tion from a man in a suit.

If you stick with the Year in Review approach, it is easy enough to accumulate anecdotal evidence that the emancipation of women was pro-

gressing apace in 1951. Truth was, when the boys came marching home, we repossessed Rosie's riveter, tossed her a trivet, and did our best to get her back in the kitchen. Irma Harding cut a strong figure—you could easily imagine her clicking briskly down a marble hall to some office somewhere—but she remained the creation of company men, and to gauge their perception of the sexes we need only consider the slogans International Harvester dreamed up to headline the 1951 advertising campaign. Tagline for trucks? *"Every model Heavy-Duty Engineered!"* Refrigerators? *"They're femineered!"*

Under threat of a compulsory bikini wax from Germaine Greer and the editors of *Bitch* magazine, let me state for the record that the term *femineered!* is a real time-warping mind-bender. The sexism is one thing, the blitheness quite another. Cheery condescension meets leering futurism. I understand now why my favorite nursing professor went all Valkyrie on me when I referred to my lab partner as a *girl* in 1987. I always thought the sisterhood of second-wave feminists who ran the University of Wisconsin at Eau Claire School of Nursing were a tad uptight, but *femineered!* goes a long way toward explaining their frame of mind.

If I had a bra, I'd burn it.

○ ○ ○ ○

My truck is an ugly truck. I picked it up in college, bought it for 150 bucks from a guy named Ron. Ron used the truck to haul firewood—it wasn't even licensed for the road. The truck body was originally a hearty red, but at some point Ron swabbed it with a coat of pink primer. He used a six-inch-wide paintbrush. There were a lot of drips and runs. In other spots the primer is spread so thin that the bristle pattern is highlighted in hairline streaks of surface rust. When he finished, Ron signed and dated his work, daubing a knockoff Playboy bunny beneath the spare-tire bracket.

The front end of the truck is blasted with rust. The grille has deteriorated to the point that the headlights wobble in their sockets. You can stick three fingers through the gaps in both front fenders. The bumper is bent. Before I parked it the last time, the radiator was blowing green mist. The front windshield is cracked in the vertical, and rain leaks around the weather stripping and streaks the dash. There is a boil the size of a grapefruit on the left front tire. The speedometer never has worked, and the deck of the bed is so riddled with holes that you could load half a yard of gravel and over just three miles of bumpy road, sift the sand from the stones. To a large extent, the truck is, as they say, shot.

On the other hand, I've never done a thing to that old six-cylinder engine beyond changing the oil, and yet it has always spun smooth as an antique safe dial. The cab doors close with a seamless click, just like they must have done when the last line worker tested them fifty-some years ago. The lugs on the back tires are fat and nasty, no wear at all. The wiper blades run off the vacuum system, so they bog some when you punch the accelerator, but then you back out of it and they whip-whip-whip as if they're making up for lost time. You could pop corn on the in-cab heater. And after six years at a dead stall in my driveway, all four tires are still holding air.

I bought the truck when I was in my final year of nursing school. For the next three years it was my sole means of transportation. I drove it to school, I drove it to work, and I drove it up north to visit my folks or go deer hunting. There were problems. The truck tended to lock up and refuse to crank when the engine was hot. If I switched it off, I'd have to wait an hour before it would start again. It was a little light in the rear end, and didn't handle so well in the ice and snow. Sometimes the headlights blinked on and off. And I used to run out of gas a lot. The gas tank was huge, but the gauge never worked. I stashed a curved stick behind the seat and took readings with that. I'd thread it down the angled gullet of the fill pipe, then draw it out and estimate my range of travel based on how much of the stick was wet. The calculations were inexact: If you were the guy who jumped out of his car in 1989 to push me through the stoplights at the intersection of Clairemont and Fairfax, thank you. Ditto the Samaritans of Hastings Way and Eddy Lane. And here's to the boys at Silver Star Ambulance Service: When I misread the stick and stalled at the troublesome convergence of Birch Street and Hastings Way, they spotted me and sent a uniformed EMT dashing through traffic to rescue me with a can of lawn mower gas.

As ugly as the truck was the day I bought it from Ron, it has only gotten worse with the years. More rust. Bigger holes. A couple more dents. One windshield wiper is missing. I was driving on the freeway in a blizzard and the blade was icing up, so I rolled down the window, reached around, and, timing the wiper—an old Wisconsin wintertime motoring trick—grabbed and snapped it every time it came past. The idea is to knock the ice loose. On the third try, I mistimed the grab and the wiper came off in my hand. I drove the rest of the way with my head out the window, squinting against the snow and dipping my head back in the cab at regular intervals to shake the ice-cream headache. The other wiper is still intact, but held in place with athletic tape.

In 1989, I took off for Europe and left the truck behind my grandpa's barn. It must have weathered the dry docking fairly well, because I remember driving it to work in the early 1990s, but sometime around 1992 it developed that radiator leak and I parked it again. In 1995, I moved north to New Auburn and had to have the truck hauled up on a flatbed. A local mechanic got it to run one more time, but the radiator still leaked, and then the gas tank rusted through, and that was that.

I am homing in on forty years old. Another twenty years and I'm looking at sixty, and these days, twenty years seems like next Tuesday. I feel young but pressed for time. I am beginning to get a sense of all I will leave undone in this life. It makes my breath go a little short. I'm not desperate, just hungry to fill the time I am allowed. To cover new ground. I could be wrong, but I don't think I'm having the vaunted midlife crisis. I'm not trying to reclaim my youth or recapture the past. I just want to get that truck running. The past belongs where it is, as it is: an essential, fault-riven foundation for the present. I don't expect that truck to take me anywhere but down the road. I don't plan to, as my cousin used to say, *cruise chicks*. Although I'd be a stone-cold liar if I said I didn't think sometimes of blowing down some back road with a woman over there on the right-hand side, her eyes turned toward the fields passing by, a hint of upturn at the corner of her mouth.

This is not going to be a restoration project. I don't want to rehab this truck only to spend my weekends with a chamois and a bottle of polish. I don't want to circle the fenders nervously, checking for dings like some pastry chef searching for hair in the cupcakes. I just want to patch a few holes and punch out a few dents. Scrub the rust down and lay on some paint. I want to get this truck running so I can, as the boys in high school used to say, go "bombin' around." Ramble off to my brother's farm and return with a load of barnyard dirt for my backyard garden. Drive up Old Highway 53 to the IGA to buy potatoes and bacon. Hammer down the swamp road in the moonlight. I want to bump down a logging trail in November and back the truck up to the gutted body of a whitetail, the cooling meat destined to feed me in the year to come.

I'll have to get with it, because I tend to sit on stuff long past the hatch-by date. Which in most cases is fine. There is enough ill-considered hastiness in the world. Trouble is, at some point you keel over and croak. The man with differently sized eyes standing in my yard asking to buy the truck—he wasn't the first. It happens two or three times a year, someone knocks on the door and wonders if the thing is for sale. They all get the same answer: *no*. I want to fix it up. The squint-eyed man looked at me like he didn't believe it would ever happen. He squinched that bad eye down extra tight, spit again, and then he got in his own truck and headed back home to Chippewa Falls, where for all I know there is an International refrigerator in every home and the streets are paved with gold.

◌ ◌ ◌ ◌

On behalf of the brand, I apologize for the whole femineering thing. It's like the big farm kid trying to fashion a homemade valentine from a beer carton using a pair of tin snips. International was never about the ineffable essence. "For many," reads the copy on the back of *International Truck Color History*, "International Harvester means handsome, sturdy, no-nonsense trucks."

Handsome. Sturdy. No-nonsense.

Irma.

Back to the *Color History:* ". . . International trucks have an earthy, wholesome quality that makes them attractive."

Earthy. Wholesome. I pull out my copy of *Freezer Fancies*. Irma's eyebrows are trim but not overplucked. Her hair is pulled back and gath-

ered, but not so tight as to eliminate the waves. A ruler-straight part
runs the center of her scalp, suggesting that Irma values tidiness and
discipline, but the loose curls bobbing above the nape of her neck imply
that fun will be allowed. The gathered hair is too informally hung to be
termed a bun. You see a little fluff and dangle. Something to shake loose.
The carefully combed part speaks of a tight ship, a neat sock drawer and
folded underwear, but the curls are there to be let down when the work
is done. I am also thrilled with the wash of gray above her left brow.
It courses backward in easy waves, and it speaks to me of experience.
Whither your sulking supermodels in the face of this bright, strong,
touch-of-gray woman! She holds her head erect on a graceful neck, she
is wearing nicely turned button earrings, and she has a dimple just below
one strong cheekbone.

But above all, I am taken by those eyes. Irma Harding is my Mona
Lisa. You will hear Mona Lisa described as enigmatic. Irma looks more
energetic than enigmatic. I look at Mona Lisa and I think, there's a girl
who would run up your credit card and pout. I look at Irma, and I think,
there's a woman who keeps her checkbook balanced. I think, there's a
woman who would be pleased to ride in a truck. Shoot, she could *drive*
the truck, and I bet she can double-clutch like a full-on gear jammer.
You look at Irma's eyes and you think, there's a woman who wouldn't
mind a little wind in her hair, a little muss. I imagine Irma riding beside
me in the truck, I'd look over there and that part would be holding but
those curls would be blown, and she'd be grinning. If those boys at International had kept Irma on board to sell trucks instead of refrigerators,
they'd still be in the thick of things. Frankly, I'm not sure she would have
put up with their guff. I look at the picture again, reconsider the lines of
her jaw and the steadiness of that gaze, and I get the feeling that if a guy
messed around with Irma, he'd wind up doing long stretches listening to
Otis Redding albums in the dark.

Back in the real world, I am long past conjuring a woman who would
even *have* me, never mind *suit* me. I simply have no idea who she might
be. I went on my first date at sixteen. Lisa Kettering. I kissed her in
the moonlight shadow of a pine tree, and she cut me loose inside of
two weeks. Now, at thirty-eight, I have a relationship track record that

can be summarized in a single overwhelming understatement: *the art of going steady eludes me*. And after two decades of having the mirror to myself, I have cultivated an accumulation of tics and idiosyncracies bound to unhinge the most long-suffering angel. Lately I've been in monk mode. No dates for nearly a year. When I met my last girlfriend, I fell headlong. Without sense or reservation. She drove a beautifully beat-up blue pickup, rolled her own cigarettes, and painted her motorcycle John Deere green. You should have seen her in a pair of work boots, backhanding the sweat from her brow. I goobered along behind her like a ridiculous balding teenager. But then came an all-too-familiar time when our conversations went dead between the lines and I got the old gut-sink. For a while I lived in hope—an eggshell kind of hope—and then one day I heard a country music song with a first line that went, *"I won't make you tell me / what I've come to understand . . ."* and I just thought, *a-yep*.

We plunge into love with a naïveté that ignores all prior humiliations. Thank goodness, I guess. Because we never learn, we reach for love again and again.

CHAPTER 2

JANUARY

THE YEAR BEGINS barren and brown. There should be snow, but the land lies stripped in subzero wind. Among the remains of last year's garden the tan stalk of a dead tomato plant ticks against the spare wooden frame that propped it through the fat green days of summer. The stalk wavers along a brief arc, dipping herky-jerky like the wand of a failing metronome. The plant yielded some good tomatoes. I roasted them in a deep pan with salt, olive oil, cloves of unshucked garlic, and sprigs of thyme. You ladle off the juice every twenty minutes or so and freeze it for a sweet, delicate stock best consumed during snowstorms. The residual pulp gathers body from the garlic and spirit from the thyme. The spent garlic, when squeezed warmly from its husk directly upon your tongue, will slacken your face and make you shimmy.

The stock and the pulp are in my little chest freezer now, down in the basement where the fuel oil furnace has been firing all day. I can hear the blower kick in, a muffled rumble followed shortly by a huff of warm air through the grate. I am on the second floor at the head of the stairs, standing at a window overlooking the garden, which is comprised mostly of raised beds—loaves of soil contained by rectangular frames constructed from two-by-twelve planks in the manner of a sandbox. The tomatoes that hung from that plant were pale yellow and big as a baby's head. I grew them from a seed harvested from a plant sown by my sister-in-law. She planted her garden in spring and didn't live to see fall, killed in a car wreck in her seventh week as my brother's wife. The tomatoes

are called Amish Yellows. We write "Sarah's Tomatoes" on the little plastic flags.

The windows in this old house are loose. The wind sets them to rattling. Looking down, I have my face close to the pane, so close I raise
a little fog and smell that cold window glass smell, the scent of ice and
dust. The wind rises and seeps past the sill. I imagine this outside air
purling and tumbling through the warm inside air the way water curls
through whiskey. My nose is cold at the window. The earth is frozen dirt.
I think of the grave.

◇ ◇ ◇ ◇

In the post office lobby, down to the Gas-N-Go, at the monthly meeting
of the fire department, everyone I meet is bemoaning the lack of snow.
Nobody—not even the natives—likes the cold, but cold and white you
can take. Snow obscures the grit and covers the trash. Snow pretties
up the scene. Renders it bearable. Whereas cold and brown leads to
drink and desolation. All around town the snowmobilers are moping,
their sleds trailered up and waiting. They drive to the bar in their pickup
trucks and pine for a blizzard so they might drive to the bar on their
snowmobiles. Even at twenty below, snow brightens the bleak earth. It
is a postcard effect, and won't do you a lick of good if you slip and break
your hip on the way to the mailbox, but it can be enough to keep you off
the sauce. On a more fundamental level, it insulates the topsoil, limit-
ing the depth of freeze. Exposed as they are, my raised beds are extra
vulnerable. I neglected to mulch them with straw last fall, and now they
are frozen through and through. The last couple of years I have been
nursing a haphazard little collection of perennials. Summer savory, some
sage, and a delicious cluster of lemon thyme. Now they are almost certainly dead forever.

I am an idiot for failing to mulch. It would have taken me all of fifteen
minutes. I'm particularly chagrined about the lemon thyme. It was a
gift from my friends John and Julie. Every year they oversee a magnificent garden. They sprouted a cutting, folded it into a moist paper towel,
sealed the packet in a baggie, and sent it to me through the mail. I got it
to take root and it thrived. By the end of summer I was using it to make

pan-roasted breast of chicken. A little olive oil, a little brown chicken stock, some pepper, and the clean lemony notes of the thyme. Simple. Delicious. Now the green is gone, the bush a sparse tangle of stems.

As a longtime bachelor it is a matter of overblown personal pride that less than ten frozen pizzas have crossed my threshold since I bought this house. Sadly, there have been other lapses. A few years back, I had some blood work done. My "bad" cholesterol was mildly elevated. If it gets any higher, my general practice doc said, we should consider medication, but for now, give it a year. Watch your diet, see if you can bring it down. In the isolation of the doctor's office I resolved to eat nothing but alfalfa sprouts and apple wedges. I braced for the pending austerity by grabbing a burger and curly fries for the drive home. For the next year and a half, I paid strict attention to my diet, consuming whole wheat, tofu, baked fish, lettuce, broccoli, all those do-gooder foods. The ones that leave you feeling dietetically righteous.

And hungry. Which is to say after all the conscientious nibbling, I would fling myself off the wagon. Follow the tofu nibbles with deep-fried cheese curds. Lay a foundation of fresh vegetable salad, then brick it over with half a tray of caramel bars. Skinless chicken followed by chocolate of any formation or quality. Carrots and half a bag of mini-doughnuts. And if the cupboard is bare, a mad four-block dash to TJ's Food-N-Fun for a Tubby Burger. Nineteen months later, I had another blood draw and found my LDL up another ten points.

So my willpower rates a big fat zero. But what a repulsive thing to associate with food: *willpower*. As if one would parse out love or oxygen by the teaspoon. When I look at my picture of Irma Harding on the cover of *Freezer Fancies,* I think, sure, she'd make me eat my spinach, but then she would slip me a batch of butter-larded freezer cookies, a basket of shredded coconut balls, or a perfectly engineered chunk of chocolate whipped-cream cake. I would eat them right down, and she would grin at me, drawing one side of her mouth back from that beautiful, tad-crooked tooth, and she would ask, *Baby? Are you still hungry?* And I would say, *Oh yes, Irma, Oh yes I am.*

My four seminal culinary influences—listed in order of appearance—are:

1. Jacques, a highly skilled and half-crazed emergency medical technician. We pulled a lot of forty-eight-hour weekend shifts together around the end of the 1980s. Our headquarters were in a funeral home, and in between ambulance calls we prepared our meals in a little kitchen ten feet from the embalming room. Jacques taught me to rub venison with allspice. Sounds simple, but it was my first exposure to red meat jazzed up with something other than Lawry's Seasoned Salt. Jacques's derring-do struck me as *très haute,* and opened my mind to further possibilities. Allspice was my gateway drug.

2. Jim Harrison and every word he's ever written about food, even though some bemoan all the garlic. More than the words, the *way.* Gusto meets reverence.

3. Jennifer Paterson and Clarissa Dickson Wright. The *Two Fat Ladies.* Brought to you by the BBC. I watch and rewatch their five-video box set. Jennifer and Clarissa empower me with broad license and anchovy paste. As an aside, Jim Harrison is on record poor-mouthing the use of butter in cooking, while the Two Fat Ladies lob it in at every turn. With the recent death of Jennifer, we have lost the chance to resolve the issue the way it should be resolved: in a three-way triple-threat steel-cage match featuring spatulas and olive oil.

4. *Think Like a Chef,* by Tom Colicchio, whose message I rightly or wrongly took to be, *regarding recipes, improvisation trumps pedantry.* Having said that and scorched the cutlets, it doesn't hurt to recall that Picasso drew a lot of orthodox ladies before cranking out *Woman in a Hat.*

Riffing off these four muses, I have concocted rosemary-rubbed venison roast served with a red wine and shallot reduction fit to make an atheist say grace, plated chilled leeks drizzled with a dill mustard vinaigrette that left me trilling aloud, and I once faked up a duck soup that I am certain was eligible for several international cooking awards. On the other hand, I have also created frankly repulsive stir-fry eggplant parmesan the consistency of oil-soaked felt, and marinara resembling bloody library paste.

When something is a success, I jot down the recipe and pin it to a three-by-four-foot bulletin board I fastened to the kitchen wall with drywall screws when I moved into the place. These recipes don't qualify as recipes in the formal sense; they're more a record of ingredients that went together well when I threw them together. Measurements, if they are cited at all, are generally denoted in increments of *glug, slosh,* or *tad*. Over the years I've lived in this house, the ingredient lists have accumulated like handbills, shingling over one another so that it often takes a minute or two of leafing and peeking before I locate the one-off venison and parsnips mélange that tasted so good last March. When one of these improvisations turns out, that's when I notice my singlehood. That's when I miss someone. You want to look up from your plate with a smile and just shake your head at the fundamental wonder of food and the civilized joy of convening to eat.

I had this moment where I thought I might try baking my way through *Freezer Fancies,* but then I checked my calendar and I will be on the road ten days this month and the dishes are stacking up as it is, so I have decided to choose one recipe and see how it turns out. This decision of course has nothing to do with my schedule and instead is predicated on my inability to stay on task, combined with a healthy respect for baking, which I distinguish from "cooking." When I cook, I tend to wing it. Baking requires follow-through and exactitude, to which I respond, Hey! Wanna go ride bike? Despite fond memories of working with teaspoons and measuring cups and learning the difference between baking powder and baking soda while making chocolate chip cookies from scratch with Mom, I rarely do any baking.

It's fun to review *Freezer Fancies*, with its vintage graphics and ver-
nacular. I'm going to take a pass on the Pink Party Cake. Ditto the Ice
Cream Bell, the Pink Tapioca Pudding, the Meringue Shells, and the
variously complicated Nest o'Balls. Nor shall I make the Ice Cream
Man, directions on page 12: *"This gay little fellow will be the life of any
party and the kiddies will love him!"* Time has a way of modulating the
lexicon. Ultimately, I chose to make the Frozen "Six-in-One" Cookies.
*"Your kitchen will develop into an after-school 'hangout' if you use this
assortment of 'melt-in-your-mouth' cookies!"* I swear I'll call the cops.

The Six-in-One recipe was simple enough, but I modified it, going
Four-In-One. I skipped the coconut and the raisin versions, and wound
up with four wads of dough: plain, chocolate, pecan, and a cinnamon-
and-nutmeg combination. I rolled each dough ball into a cylinder one
and three-quarter inches in diameter, sliced a few cookies off for immediate baking, wrapped the remaining cylinders in plastic, and placed
them in my freezer. I put the cookies in the oven and brewed coffee.

Until I came across *Freezer Fancies* and set out to collect Irma's entire
oeuvre, I was in possession of exactly thirteen cookbooks. A comparatively
modest collection, but I have my reasons, the main one being, nothing
snarls me up like options. I blame this on my genes and my waste-not,
want-not penny-pinching proto-Calvinist roots, which imbued me with
the feeling that to be in possession of a useful thing and not use it is to
allow the devil to wedge his big toe in the screen door of your soul. This
line of thinking engenders teetering stacks of hand-washed yogurt cups,
bales of folded grocery bags, impassable porches, and the hoarding of
broken-handled snow shovels. It follows that the implied responsibility
inherent in a collection of cookbooks is overwhelming.

Genetically, the problem is exacerbated by the fact that I am pathologically unable to maintain the process of linear thought. In conversation I rarely plow halfway through a sentence before my attention,
best characterized as a dim-witted antelope, spots a flag waving from the
topical periphery and skips off to investigate. My brain zigzags like an
amped-up puppy bounding around the chicken yard, never able to pick
just one bird and stick with it. It's like *Rain Man* in here, minus the apti-

tude for math. I have a hard time starting things and I have a hard time finishing things. Once while speaking at a camp for troubled youths on the topic of how they might get their lives together, I looked down at my sandaled feet and noted I had trimmed nine out of ten toenails.

Combine guilt-ridden sense of duty with terminal indecision and you will understand why I resist bringing any more cookbooks into the house. I look at my stack of thirteen, and I hear an austere Depression-era voice in my head, saying, *Hundreds of perfectly good recipes in there, and you haven't even* touched *them*. There is work to be done, and I am *way* behind. I've tallied the workload, and it freaks me out:

Betty Crocker's Cooking for One. A gift from Mom when I got my first apartment. Reminds me of the love we never sufficiently return and thus the very sight of it renders me melancholy. Tears in the spaghetti sauce. Leafing through it now, I note that I have yet to compose Chicken Livers in Toast Cup. Number of recipes: 177.

Untitled. A pamphlet of recipes published by the China Village company and given away as a premium for the purchase of a wok. Got the pamphlet when Mom gave me the wok. Used the wok a lot, but not the recipes. I note my mother has inscribed *"bland"* beside Chicken with Mushrooms. This from a woman who wears her hair in a bun and once eased the family through a lean stretch by feeding us boiled wheat from a plastic trash can. Her idea of *bland* implies an absence of flavor so utter as to create a vacuum capable of bending light. Number of recipes: 11.

Kenmore Microwave Oven Use & Care Manual and Cookbook. Never used it. Number of recipes: 25.

Simple Cuisine, by Jean-Georges Vongerichten. I bought the book after reading a review extolling its combination of simplicity and sophistication, principles I trust guided Jean-Georges when he catered Donald Trump's third wedding, where simplicity and sophistication convened for Foie Gras with Quince-Pineapple Compote, Lobster Daikon Rolls

with Rosemary Ginger Sauce, and Caviar-Filled Beggars' Purses Topped
with Gold Leaf. I have memorized his edict that the essence of any vinaigrette is to use one part acid to two parts oil. Number of recipes: 205.

Balti Curry Cookbook, by Pat Chapman, founder of the Curry Club
and self-described "curryholic." Sometime in 1984 at roughly 1:15 A.M.
Greenwich Mean Time my English friend Tim stumbled from a Can-
nock pub and led me to an Indian restaurant called the Padma, where
I ate my first curry. I have craved coriander and poppadams ever since,
but stop short of calling myself a "curryholic." When Tim mailed the
cookbook, he enclosed a few Curry Club packets with which I was able
to concoct some passable dishes. As for the cookbook, I admit I have
never used it, in spite of the titillation inherent in a subtitle that promises
to reveal *"The exciting new curry technique."* Number of recipes: 100.

Indian Meat and Fish Cookery, by Jack Santa Maria. Also from Tim.
This one I used, if only to make my own Garam Masala, which sadly
came down on the side of sawdust. Number of recipes: 239.

Beaver Tails & Dorsal Fins, by G. Lamont Burley. Subtitle: *Wild Meat
Recipes.* My rifle-toting grandfather gifted my nonhunting mother with
this little number during the deer hunting season of 1986, which by
chance coincides with that period of time in which my younger brother
John was working through his amateur taxidermist phase and had become
prone to storing partially resurrected subjects in the freezer. You'd get a
hankering for some maple nut ice cream and find your access blocked by
pelts and frozen snoots. I retain the book for sentimental reasons, and
for the possibility that I may one day be required to barbecue a skunk
(page 16). G. Lamont Burley claims to be an all-around woodsman and
ridge runner, and I believe him. Number of recipes: 42.

Bull Cook and Authentic Historical Recipes and Practices, by George
Leonard Herter and Berthe E. Herter. In the tradition of G. Lamont
Burley, only edgier. Consider the introduction: *"For your convenience
I will start with meats, fish, eggs, soups and sauces, sandwiches, vegeta-*

bles, *the art of French frying, desserts, how to dress game, how to properly sharpen a knife, how to make wines and beer, how to make French soap, what to do in case of hydrogen or cobalt bomb attack. Keeping as much in alphabetical order as possible.*" Included are instructions on how to buy wieners. Number of recipes: 94, not counting those for beer, wine, and French soap.

Rival Crock-Pot 3½ Quart Stoneware Slow Cooker. Recipe pamphlet for use with the Model 3100. The Beef Stew is not bad. I cannot vouch for the Magic Meat Loaf. Number of recipes: 25.

Betty Crocker's Dinner for Two Cookbook. Mom again. "You can always freeze the other half," she'd say, but the implication remained. This is the 1958 edition, and I cherish it for the funky artwork. When I was a child I leafed through it like a picture book. Number of recipes: 491.

Good Housekeeping's Family Favourites. The Great Britain edition, as you might infer from the *u.* A gift to my mother from her English pen pal Pat in 1957. Twenty-seven years later I would slink into Pat's house at 4 A.M. smelling of curry. Recipes include Jam Roly-Poly, Mutton Broth calling for scrag end of lamb, an imponderable Sheep's Head Broth, and the legendary—you're the naughty one here, not the British—Spotted Dick. Number of recipes: 500.

Let's Start to Cook. Published by the *Farm Journal.* Time and time again, I turn to this one for the basics. Most remarkable for the cover art, produced in 1966 by the design firm Kramer, Miller, Lombden, Glassman and featuring stylized green beans, cupcakes iced with what appears to be Gillette Foamy, and an orange sherbet salad the size of your head— all on a hot pink background. Ten years later the same firm designed the cover art for the Messianic Records release *Songs for the Flock,* available at press time through the Jews for Jesus Web site. *Songs for the Flock* features a photograph of yearling lambs, thus making the album cover more appetizing than the cookbook cover. Number of recipes: 300.

Think Like a Chef, by Tom Colicchio. This is where I learned to roast
the tomatoes. It is also where I picked up the term *pan-roasted,* which
sounds simple and classy, even if you're just frying chicken. The photographs in this cookbook are pure glistening titillation. Number of recipes: 111.

Thirteen cookbooks, 2,320 recipes. And you wonder why I get short
of breath? Not so bad if you cook three meals a day. But right off the bat,
you figure breakfast is shot, recipe-wise. I make pancakes maybe twice
a year, usually from the recipe on the mix box. Otherwise it's coffee 'til
noon and whatever carbohydrates I can scrounge, the primary danger
being that I will run an errand taking me within six miles of a gas station with a doughnut rack. Lunch—even if you work at home, which I
mostly do—is rarely the time to cook. Usually I eat out of a can or plastic
container, or make a sandwich. So let's say I'm very rigorous and use
three—no let's not be silly, *two*—recipes to make supper. Subtract for
the fact that I wind up eating outside the house two nights per week.
Take off two more nights for all the times I wait until I'm too hungry to
cook properly and instead binge on fig newtons, jerky sticks, or a 1992
Minnesota Twins plastic stadium cup full of Lucky Charms. Now we're
down to six recipes per week. I am blessed with charitable friends who
invite me in (or are too polite to turn me away) for dinner roughly twice
per month, and as long as I remain on speaking terms with the relatives, Christmas and Thanksgiving are off the table. Ten days a year I go
deer hunting dawn to dusk and return home sapped of culinary initia-
tive. Once a month I shop hungry and wolf down broasted chicken and
jo-jos in the IGA parking lot. Finally, the average adult has two to four
common colds a year, average duration one to two weeks, let's split the
difference and say three ten-day bouts. No one wants to cook food they
can't taste, so that's another thirty days shot. At this optimistic pace it
will take me fourteen years, ten months, and fifteen days to get through
every recipe in the house.

And we haven't even addressed the Internet, otherwise known as the
Devil's Mind-Fryer. I recently developed a jones for snickerdoodles, so
I entered "snickerdoodle recipe" into a search engine, which in point-

three-eight seconds returned the usual thousand hits and assorted iniquities. The world is impossibly ornate. Feeling a twinge of panic at all the overload, I selected a single recipe site, the plan being to narrow things down. I typed "snickerdoodle" on the home page and it scrolled out twenty recipes. *Twenty recipes . . .* for a cookie containing a sum total of seven ingredients not counting the cinnamon. Operating in this range of abundance locks me up. How in the name of sifted sugar do you choose? What if the all-time world-record blue-ribbon who's-your-grandma finest snickerdoodle recipe ever committed to a gingham-trimmed note card is sitting there like one of the nine original Beanie Babies at a yard sale and I skip it for a mistranscribed abridgement of Aunt Tooty's Double-Doodle Snickerdoodles supplied by Sylvia G. in Omaha who frankly skimps on the butter? How will I know what might have been? The logistics and bulk staples required to cook one batch of all twenty recipes are prohibitive. Three minutes ago I wanted a cookie. Now I am leaning into my computer screen, hand on mouse, face frozen in a rictus of dither.

When I took them from the oven, the chocolate cookies were chocolatey to the point of muddiness and crumbled at the lightest touch—I may have miscalculated the exact dimensions of "½ square bitter chocolate." The others turned out beautifully, each bite commencing with crispy resistance, then yielding to a moist center, the richness of the butter pressing fatly against the tongue. And then a sip of coffee, the cookie sweetness melting and giving way to the dark surge of the beans.

The recipe indicated that the remaining dough could be stored for up to one year.

It didn't last three days.

In part to mitigate the barren state of the earth, I have decided to order seeds for my garden. I possess the perfect armchair for the task, a saggy old green thing that came from my grandmother's basement and now sits on a rug beside my homemade bookshelves. Sinking into the worn cushions, I spend the remainder of the afternoon leafing through seed catalogs and recharging my chamomile tea. It is as if a sunlamp has been turned toward my soul. My winterbound spirit thaws, releasing sense memories—the *shink, shink* sound of a hoe cleaving sandy soil, the press of a hard seed between the pad of thumb and forefinger, the scratchy hiss of squash leaves moving in a warm breeze. I am *this close* to writing a poem. Seed catalogs are responsible for more unfulfilled fantasies than Enron and *Playboy* combined.

Blissful though it is, the annual seed catalog review adds up to a perennial tradition of willful delusion. It begins responsibly enough. Scientific approach and rigorous intent. As, for example, in the selection of beets: Notepad at hand, I calculate the harvest date of fifty-three-day Red Aces as opposed to sixty-day Cylindras, factor in the hybrid vigor of the Red Ace, take into account the sliceability of the Cylindra, cross-reference all results with the applicable hardiness zone, jot my selection in neat preruled columns including item name, associated catalog number, and miscellaneous starred comments, and then move briskly on to a hard-eyed evaluation of kale. I am in essence a minor god, with plans for my few square feet of the earth. I shall sow, and I shall reap. I am a catalyst in the cycle of life. I am also distracted by all the pretty pictures.

The seed catalog is printed on paper of the same texture as your gaudier supermarket tabloids—a stock perfectly suited for oversaturated photos of Royal Burgundy Purple Pod Bush beans, overfluffed sheaves of savoyed spinach, and lurid tomato shots with every fruity globe so taut and flawless it might have been snatched from the chest of a prefab starlet. Carrots are arranged in arresting bolts of orange. A neon splay of Bright Lights Swiss chard vibrates like a beer sign in a health food store. Purpling stems of beet green plunge into the dusty lavender crown of the stout root, sliced open in one photograph to reveal a glistening fine-grained core the color of deoxygenated blood. The play of sun and shadow on a grapelike cluster of Sweet Millions miniature tomatoes is so mustily conveyed that your parotids clench at the thought of the skin popping under the pressure of your molars and the subsequent sweet gush of pulp. A pair of Bell Boy peppers reflect the light with a blue tinge that suggests the exact feel of the cool green lobes against your palm, and I am drawn straight into summer. It is as if the catalog ink is spiked with chlorophyll. Rigorous intent begins to fray. Never shop for groceries on an empty stomach, they say. Corollary riff: Never order seeds when the world is frozen stiff and leafless.

Scientific process? That vaporizes the minute I hit the cucumber page. Ain't no such thing as *a* cucumber. You've got your Sweet Slice Hybrid. Your Fanfare. The Ashley. The Marketmore, the Cool Breeze, the County Fair Hybrid, the Orient Express, and the Sweet Success. The Diva. The Homemade Pickles is slotted just below the SMR–58, a juxtaposition implying a genetic journey from Grandma's backyard patch to a petri dish in some lab. Claims are made regarding the resistance of certain cucumbers to scab and mosaic. Others tolerate powdery and downy mildews. Some are parthenocarpic: able to set perfect fruit without cross-pollination. Some are designed to grow in a lowly pot, others thrive on a trellis. Over sixteen variations on a cucumber in the space of a single page. My carefully notated columns begin to dwindle.

In the end my order includes kale, carrots, parsley, dill, cilantro, summer savory, lemon balm, basil, sweet marjoram, oregano, lettuce, okra, parsnips, peas, squash, and tomatoes. Also three packets of cucumbers: a pickler, a slicer, and the Orient Express Hybrid. Checking my notes, I see I chose the Orient Express because the catalog copy said it would thrive on a trellis, and *I have a trellis*.

The seed catalogs promote several varieties of "burpless" cucumbers. I have yet to find one promoted as "burp-*ish*." This is flatly a missed marketing opportunity. Among my rural and roughneck acquaintances are no small number of folks who not only savor the art of eructation, they cultivate it. There are guys on the fire department capable of melisma. I have seen a woman throw her head back beneath the Jamboree Days beer tent and let loose a burp so resonant polka dancers were moved to applause. I know men longing to belch a full-length version of "Free Bird." Beer works, but it impairs your ability to play air guitar. There are people out here who would go out of their way to plant row on row of Burp-Mor Hybrids, County Fair Honkers, and Belching Divas.

Now that the seeds have been ordered, I have hit the apogee of my gardening season. You lick the envelope, or click *Send,* and you think, "There." As if you have returned the hoe to the shed, or bundled the last cluster of garlic. Today I am buoyed by hope and visions of a rank harvest. Once the seeds arrive, all subsequent horticulture is executed within the context of reality and is therefore trying.

$$\circ \ \circ \ \circ \ \circ$$

The snow starts late one afternoon. The first fat flakes drift aimlessly. I walk out to stand in the backyard, where I can hear the papery *tic-tic* of individual crystals striking the crisp maple leaves. The snowfall is more urgent now, and soon the ground is blurry brown. Then it is white. By dusk, the snow is accumulating depth. When the nine o'clock siren sounds from the water tower across the tracks, I step to the front porch and survey Main Street. At each streetlight, flakes drop through the mercury-vapor nimbus like moths in free fall. The village is muffled in snow. Every sharp line is softened and the windows up and down the street glow warm and yellow. In the deepening snow, even the meanest home looks cozy. I go to bed and roll up in blankets. The plaster walls are cold. Drifting, I offer a prayer of thanks that in all of time and space I have been delivered to this ephemeral cocoon, a pinpoint of warmth in the unknowable universe.

In the morning the snow is knee-deep and the temperature has dropped below zero. On Moose Country Radio, the gap between George Jones and Loretta Lynn songs grows wider as the announcer recites an expanding list of postponements and cancellations. School has been shut down. Several basketball games have been rescheduled. Over at the turkey factory in Barron, the evisceration team is starting two hours late. Outside, the morning is filled with the sound of snowblowers and the flat *scrape, scrape* of snow shovels. As we dig out, we greet one another with mittened waves and puffs of breath, cheerful as kids playing hooky. We lean to our shovels with stoic determination, secretly delighted that in the age of heated seats and convenience-store cappuccino we can still pretend to be pioneers as we strike out for milk and eggs up the block at the Gas-N-Go. The air is sharp with cold. With every inhalation, our nose hairs snap together like magnets and freeze. They thaw and separate on the exhale. We tromp around in our big boots, imagining we survive on pemmican and hardtack. The illusion doesn't last. The plows are out, and by midmorning the four-lane is whooshing with people who dared not risk the deadly trip to work or school, but now, given a day off, will drive forty miles to the mall.

After shoveling snow, I am hungry. This is the kind of day when you'd like to step through the door, stomp the snow from your feet, and inhale a hearty dinner. Sit right down and eat roast beef. Tuck into real mashed potatoes and fatty brown gravy. Savor the overcooked carrots and onions, have another slice of meat smeared with horseradish. I am content in the bachelor life, but at moments like this, I admit to old-fashioned sexist longing. Sometimes I cook up comfort food, but cooking your own comfort food is akin to scratching your own back. Same sensation, less watts. In the basement I rummage around my little chest freezer until I uncover a plastic-sealed lump of homemade pesto. I place the lump in a sauce pan over low heat and pull a pasta pan from the rack above the kitchen window. Out in the backyard, the raised beds are cloaked in snow. They look like gravestones dipped in almond bark. The row-on-row arrangement of the beds reinforces the graveyard image, and Sarah's tomato plant is a skeletal bouquet. Now I think of my brother, alone in his house those days after Sarah died. When I turn back to the stove the pesto lump is half-thawed and the fragrance rising from the pan is pure green summer.

I eat my lunch in the saggy armchair. From here I can see my old truck in the driveway. The body is dolloped with snow lumps that mimic and distort the underlying contours of the fenders and roof. I really don't know where to begin with that thing. I know I can't do it myself, but I don't want to just turn it over to someone. Besides, I can't afford to go that route. I need someone who will let me contribute a little sweat equity. Someone who doesn't mind my company. I've been using all the research, all the gathering of manuals and cookbooks as a sort of throat-clearing exercise. Now I'm looking out the window at the snowbound hulk thinking so many of my projects start off big and then languish in disarray. I need help, that's for sure. I'll have to start asking around. My mechanical abilities dwindle just past lifting the hood. Righty-tighty, lefty-loosey, and after that I've pretty much exhausted my options.

The pesto and angel hair are warm in the bowl on my lap, the fragrances of olive oil and basil blending the exotic and the familiar, equal parts sunny Tuscan hillside and hometown dirt. A meal like this makes you want to live forever, if only for the scent of warm pesto in January. When I finish the food, I'll place the bowl on the floor, shrug into the broken cushions, and doze.

My brother Jed walked a black path after Sarah died. There were stretches where we feared he might give up. We both serve on the local volunteer fire department, and sometimes an hour after the hoses were hung to dry we would still be leaning against his pickup while he talked against the darkness, holding out against returning to the empty bed. He threw his sleeping bag in his pickup and drove across the country to California and then came back. He put a lot of tears on our mother's shoulder. It got worse in the winter. He just wanted to sleep. My brother John took to prying Jed from bed and force-marching him to the woods. Jed being in no condition to run logging equipment, John left him to tend the stove in the portable shack at the timber landing. No hugging or gnashing of teeth, just a refusal to let Jed go blind in the cave. This is the kind of strap-steel love overlooked by those who misconstrue stoicism as failure to engage. In the end, the broken circle closed beautifully: Sarah's mother came to Jed one night and said there was a woman he should meet. Her name was Leanne. It worked out, and they were married.

It feels presumptuous to say anything more. Jed is a private man, and in his gaze linger vestiges of things I cannot imagine. I will end with this: Leanne took up Sarah's garden, and when I drove by their isolated farmhouse in the wee hours of night last spring, there on the porch hung a light, glowing just above a flat of seedlings. I took it as a sign that he was back among us.

CHAPTER 3

FEBRUARY

BACK WHEN I WAS still living in the city and working split shifts at the hospital, I left my apartment one spring morning to discover my truck had been tagged with a pale green parking ticket stapled to a business-sized envelope. The ticket threw me off stride from the get-go because the truck was parked well off the street in its proper numbered spot on the apartment complex lot, but it was sure enough the real deal, one of those miniature self-sealers emblazoned CITY OF EAU CLAIRE PARKING VIOLATION and appended with clusters of red-letter fine print: *Make your remittance payable to the City Treasurer . . . IF NOT PAID EXTRA PENALTIES WILL ACCRUE . . . The State Dept. of Transportation will suspend your vehicle registration.* The two lines reserved for "VIOLATION DESCRIPTION" had been filled out in hand-printed all-capital block letters:

ABANDONED VEHICLE
PUBLIC EYESORE

My ears did a hot flash. I tore open the business envelope and began to read the letter within. By the second line, I was making squeaky noises like the ones you hear when you bump a wall full of bats. By the third line, I had a full-blown case of the fuming huffies.

Alpine Managment
281 Grant St.
Eau Claire, WI.
(715) 555–1433

Occupant/Vehicle Owner

*This corespondance is in reguards to your vehicle. We, here at
Alpine have been receiving many complaints with respect to
your truck.*

   *As I'm sure you would agree that a clean and respectable
place to live is an important item to all those who reside in this
particular apartment complex. We have contacted the local
police department, as well as our own lawjers, and have found it
well within our rights to impose certain standard on our tenants.*

Now I was sputtering like a cat stricken with galloping hairballs.

   *I'm am sure that this problem can be quickly remedied.
Please call us before any more actions are taken.*

   *The fine you have already received can and will be paid
by our financial service if you resond to us with in the next 48
hours.*

   *CALL WITHIN 48 HOURS BEFORE ANY FURTHER AC-
TIONS ARE TAKEN.*

Sincerely,
E. Thomas Packard
Regional Operations Manager

   There commenced an epic snit. Independent observers would later
report the manifestation of visible indignation vapors, which for the
record are off-yellow and shoot mainly from the ears. My heart was beating high in my chest and I was quivering with pique. A second read-
through of the letter left me flat-out barking.

"We, here at Alpine . . ." We, here?!? The royal *We?!* Tone-wise, E. Thomas, we are off on the wrong foot.

". . . have been receiving many complaints with respect to your truck." I raised my green-eyed gaze to the buildings around me. Suddenly every window had twitchy curtains.

"As I'm sure you would agree . . ." You don't get to be Regional Operations Manager without knowing how to blow the twin smoke rings of insincerity and unction up the backside of those you despise.

". . . a clean and respectable place to live is an important item to all those who reside in this particular apartment complex." More than you know, E. Thomas, more than you know. My particular apartment complex window faces the pallet-stacked hindquarters of a Shopko. Specifically, the litter-snagging loading dock where a depressed woman chose to asphyxiate herself beneath her car one recent frozen morning. What manner of clunker-parking clod would corrupt such a vista?

"We . . . have found it well within our rights to impose certain standard . . ." Under which purview Alpine Management has crammed twelve hulking faux chalet apartment buildings into a single city block, half of them overlooking the service entrances to a strip mall.

"I'm am sure that this problem can be quickly remedied." Yes. I'm getting the rifle now and will be up the water tower shortly.

"Please call us before any more actions are taken."

I was jabbering with rage.

You don't work yourself into this sort of state so that you can get put on hold and blow a vein. I fired up the International and dropped the hammer for 281 Grant Street.

○ ○ ○ ○

Even when it was shiny and new, the International L-120 wasn't set up to win beauty contests. It had a squatness. The fender lines were too square. In his book *International Trucks*, author and International expert Frederick Crismon refers to the L-Series as looking "squashed." Alongside its Ford and Chevy contemporaries, the L-120 was the plain girl with thick ankles. *Heavy-duty engineered* indeed. In contrast to the refrigeration division's frothy *femineering* superficialities, the L-Series

trucks had been completely reworked when they were introduced in
1949. International advertised the trucks as "new from bumper to taillight," and according to the *International Truck Color History,* there was
substance behind the hype. Nearly everything—the engine, the chassis,
the suspension, the brakes, and the look of the trucks themselves—had
been redesigned. In fact, apart from one optional three-speed transmission, Ertel and Brownell claim that the only surviving items traceable to
preceding models were the hubcaps.

Advertisements and sales literature from the time reveal that International was especially proud of the L-Series' redesigned "Comfo-Vision
Cab." At seventy inches wide, the Comfo-Vision cab represented an expansion of ten inches over the previous K-model cab, and was pitched as
the "*roomiest on the road."* It featured an adjustable seat (previous seats
had been fixed in one position) designed to provide "*head room, elbow
room, and leg room for the biggest driver in the business*" and "*lounge
chair seating for three."* In a promotional photo of the cab taken from
the perspective of the hood ornament, two undersized men in milkman
caps gaze at the camera like a pair of bemused bookends. You could
squeeze between them maybe one installment of *The Bobbsey Twins.*
Additional selling points included the addition of adjustable vent windows (*"open and close with the flick of a thumb"*), an adjustable cowl
ventilator to admit fresh air, and live-rubber cab mounts upon which the
cab was purported to "*float."* I can report from the seat of my pants that
this claim was optimistic in the extreme.

The most distinguishing feature of the Comfo-Vision cab was a one-
piece "Sweepsight" windshield, "*scientifically curved to minimize eye-
strain and reduce glare,"* while improving "*see-ability."* In truth, the
curved one-piece windshield really was a pretty big deal, unheard of in
trucks and otherwise available in only a few luxury cars. The Sweepsight
was complemented with another distinguished first: twin rear windows
to "*promote driver efficiency, safety, and peace of mind."*

❋ ❋ ❋ ❋

The thing that put me over the edge with that E. Thomas Packard letter
was him signing off on all those typos. It is bad enough to be at the

receiving end of a head-patting lecture, but to endure a misspelled supercilious browbeating on the subject of aesthetics is beyond the pale. During the drive, I polished my speech. I intended to frame my objections in terms of the First Amendment, the *Kelley Blue Book*, and my paid-up certificate of title. I would furthermore bolster my tirade with citations drawn from city code and zoning regulations, *B.J. and the Bear* scripts, and recitations from the collected hits of Red Sovine. I would conclude with a snide reference to my having made the drive over in an "abandoned vehicle." By the time I hit Grant Street, my scientifically curved Sweepsight was spittle-flecked, but I had achieved the clarity of an assassin.

Then I couldn't find the address.

Grant Street is lined with ranch houses and bungalows. I remember thinking it strange that the rental offices were located in a residential neighborhood, and as I cruised up and down the street, looking in vain for 281 Grant amid the basketball hoops and hedges, I felt the righteous umbrage rise again. *What kind of ridiculous knothead puts the wrong address on his letterhead?* After my third fruitless pass through the 200 block, I crashed the gears and roared home. Someone was going to get a very nasty phone call indeed.

○ ○ ○ ○

Despite the emphasis on comfort and esthetics, the 1951 International was designed primarily for work. Or, more specifically, for *men* who worked. In nearly all the photographs supporting the International truck ads of 1951, the men are lifting something. Feed bags, milk cans, hay bales. If they aren't lifting, they are at the wheel, driving with their hands at 10 and 2. And if they are not at the wheel, they are under the hood. The men look sturdy and earnest. Like after-shaved farmers at church, or the guy at the hardware store who can help you with your drain trap. These are men who never leave home without a jackknife and miniature tape measure. They can do math in their head and know how many square rods make an acre. These men are fundamentally useful.

I have come across only one International Harvester whose driver appears to be getting by on his looks. He is on the cover of the twenty-page *International Light-Duty Series* sales brochure. In the picture he is pulling out of a residential driveway at the wheel of an L-120 just like mine, only yellow. He appears to be in California or perhaps Arizona. The property is lined with palm trees, and the house in the background is flat-roofed and architecturally hip. Two women stand on the lawn at a fair remove. They are slim and leggy, and wearing beautifully cut calf-length skirts. From their position and line of sight you can tell they are eyeing the driver. His face is partially in shadow and he is wearing mirrored aviator sunglasses. His white T-shirt is tight across the chest. His cheeks and chin are stubbled. He looks timelessly cool. If James Dean had driven a three-quarter-ton International, he would have made it to Salinas.

○ ○ ○ ○

When I returned from my Grant Street goose chase, I had a rare moment of lucidity. Perhaps Grant was one of those streets interrupted by the river, or a railroad. I may have been on West Grant Street when I should have been on East Grant Street, or vice versa. I unfolded the city map bound in the local telephone book. Grant Street ran uninterrupted from end to end. Certain now that the rental company had flubbed their own address, I dialed the phone number on the letterhead. Probably flubbed that, too, I thought. The dial pattern seemed vaguely familiar. I was trying to place it when a woman answered.

"Hello?"

"Alpine Management?"

"Excuse me?"

"Alpine Management?

"I'm sorry, you have the wrong number."

The *incompetence*!

"Is this 555-1433?"

"Yes . . ."

And then I placed it. Eric Teanecker, a friend from high school. We lost touch, then wound up working together at a roller-skating rink during our college years. He had gotten married. I hadn't seen him for over a year. He used to tease me about the truck. This is the number I used to call to see if we could trade shifts.

"Ahh . . . is this Renée?" His wife.

"Yes."

"Eric there?"

"No . . ."

"I'll call back."

○ ○ ○ ○

All those marvelous pictures advertising the L-Line, and not one woman at the wheel. When they do appear, it is in the background, where they elbow-tote their purses and chatter with each other, or look on admiringly as the men lift things. Chauvinism aside, when it came to women and trucks, International really missed the boat.

I know exactly how they feel. In 1992, I was traveling to Black River Falls, Wisconsin, in my reliable-if-not-zippy four-door '78 Impala when the radiator blew. It could have been worse, as I was rolling up the exit ramp at the time, and had sufficient momentum to reach the Wal-Mart parking lot. (As quick as we are with the *why me, why NOW?* rap, it seems a matter of karmic responsibility to acknowledge those instances when bad luck has good timing.) I was en route to research a magazine piece on canoeing, and had agreed to meet a local guide at a downtown hardware store, so I left the car astride its expanding green puddle and walked the rest of the way.

The guide—a petite young woman named Cindy who kept her blond hair pulled into a ponytail with a pink scrunchie but carried herself with a trace of jock swagger—determined immediately that I wasn't the kind of guy who could fix his own radiator, and arranged to have the Impala towed to a local shop. Then she loaded me into her pickup truck. The exterior was bashed and scuffed, and the cab was awash in good working-class trash—spark plug boxes and empty gasket packs, shell casings, that sort of thing. We accelerated manfully from the curb. She handled the stick shift with authority, which gave me certain twinges. And I admit I noticed how her quadriceps arched against her shorts when she worked the clutch. She told me about her motorcycle. When we reached the landing, she jumped out and didn't wait around for me to help lug the canoe. Once launched, we floated the tannin-stained Black River for a

long while. Cindy paddled smoothly and pointed out key fishing spots.
"I do a lot of bass fishing," she said. *All this, and a pickup truck,* I kept
thinking. When she drew my attention to a specific cluster of brush and
identified it as the spot where she shot her biggest buck ever, I decided
it was time to get married.   ·

It's tough to make a marriage proposal in a canoe. I'm not saying it
can't be done, it's just that canoes are notoriously tippy, and because I
have never mastered the j-stroke required to successfully captain a canoe
without switching the paddle side to side every two strokes like an inde-
cisive milkmaid assigned one dasher and two churns, I had been quite
rightly placed in the position of emasculation: up front facing forward,
where I could paddle away mindlessly without yawing us madly into the
tagalders. As I recall, I got Cindy to sit in the front of the canoe only long
enough for me to get a photograph of her gazing downstream. I found it
difficult to focus on anything other than her exposed shoulders.

What I had in mind was a sandbar. With a deft variation of her stroke,
Cindy would put us ashore. Kneeling shoulder to shoulder on the beach,
we would coax up a fire with moss and flint, then roast frog-legs-and-
cattail-root shish kebabs over driftwood coals. Later, while loitering in a
muskrat-pelt loincloth and waiting for the tin-can coffee to boil, I would
pop the question. The evening would culminate with a postprandial
arm-wrestling match, loser wears the engagement ring. In the morning I
would take her in my arms and bear her to the canoe. Or vice versa. We
would emerge from the wilderness to notify our friends and reserve the
Legion Hall.

The fantasy broke when a well-muscled and wholly corporeal local
boy hailed Cindy from shore. There followed a good-natured exchange
of insults that implied familiarity. He looked woodsy and capable. A real
back-of-the-canoe fellow. I was building him up in my own mind. Then,
out of the corner of her mouth, Cindy said, "He's a weak-tit." One feels
the gonads shrink. I hope it is a sign of progress when a man subverts
machismo to allow room for frank self-assessment. Unable to construct
a scenario—beyond faking a seizure and flipping the canoe—in which
I would leave Cindy breathless, I resumed my brute-force paddling.
Shortly we debarked. Cindy dropped me at the shop with my Impala

and I have not seen her since. I recall her shoulders in the sun and the flex of her calf when she hit the gas.

○ ○ ○ ○

It isn't just the idea of a woman in a truck. At this point, they're everywhere. The statisticians tell us today's woman is as likely to buy a truck as a minivan. One cheers the suffrage, but the effect is dilutive. My head doesn't snap around the way it used to. Ignoring for the moment that my head (or the gray hairs upon it) may be the problem, I think it's not about *women* in trucks, it's about *certain* women in *certain* trucks. Not so long ago I was fueling my lame tan sedan at the Gas-N-Go when a woman roared across the lot in a dusty pickup and pulled up to park by the yellow cage in which they lock up the LP bottles. She dismounted wearing scuffed boots and dirty jeans and a T-shirt that was overwashed and faded, and at the very sight of her I made an involuntary noise that went, approximately, *ohf . . . !* I suppose *ohf . . . !* reflects as poorly on my character as a wolf whistle, but I swear it escaped without premeditation. Strictly a spinal reflex. (My friend Frank once walked around a street corner and came face-to-face with a woman so stunning he yelped, *"Jesus Christ!"* This from a poet and charter member of the local Student Feminist Alliance.) The woman plucking her eyebrows in the vanity mirror of her waxed F-150 Lariat does not elicit the reflex. Even less so if her payload includes soccer gear or nothing at all. That woman at the Gas-N-Go? I checked the back of her truck.

Hay bales and a coon dog crate.

Ohf . . . !

○ ○ ○ ○

Here lately I have been pondering the commodification of higher consciousness as evidenced by the fact that you can get a yoga mat at Wal-Mart. The world needs all the Sun Salutations it can get, but when I leaf through the glossy yoga magazines, I want to know where all the stiff and lumpy people are. (And a gentle pox on yogis who insist on taking out ads in which they pose as human origami. Gratuitous convolutions are to inner peace as home run contests are to baseball. I keep thinking of the little kid who flips his eyelids inside out and does a little monkey dance hoping you'll notice.) Spiritual discipline, shined up and streamlined for that dream focus group where census and disposable income intersect. If you were there in the beginning—if you did yoga prior to the advent of monogrammed zippered mat bags—you feel a little peevish.

Different crowd, but do you remember when they did it to cigars? My buddy Al is a connoisseur of small-town taverns, smoked carp, and cigars. He'll burn a cigar on his own, but he most of all enjoys smoking in the company of old-timers of the sort who wear stained T-shirts and go bobber fishing down by the bridge. Guys who *chomp* as much as smoke their cigars. I met Al in the early 1990s, right about the time cigars boomed. Bill Cosby and Jack Nicholson were on the covers of *Cigar Aficionado,* as were Demi Moore and supermodel Linda Evangelista. Rush Limbaugh and Bill Clinton were popularizing the cigar across lines of politics and propriety. But perhaps most irritating to Al and his old-school pals, cigars became popular with droves of lean women, high-fiving frat boys, and young sharks in suits. The anachronistic recalcitrance of their habit was suddenly *happening*. To be seen smoking was to be seen as to be trying to keep pace with the It Girls and Boys. Al defined the problem as "Goddamn yuppies."

The world of American culture and commerce functions like a combination of sponge and sandpaper, absorbing everything and smoothing it down so it slides easily into a designer shopping bag. It's the American free enterprise system at work, and while in general I am a fan, I admit to some grumping while I try to work out exactly where it is that egalitarianism gets tromped by commodification. At what point does the genuine article become frothy? I'm pretty much a live-and-let-live agnostic, but whenever I see churches luring people to their services with puppets and guitars, or these mall churches where they park your car and serve you lattes and let you watch the pastor on your choice of five JumboTrons, I want to say, No, No, *No*. Church should not be easy. Church should be *hard*. I have read that in his last days, Jesus Christ fell on his face and sweated blood. The least you can do is sit on a hard pew and squirm some.

Harleys, tattoos, party platforms, the Wild West, we distill the concept to its iconic essence, slap a price tag on it, and get down to the business of overexposure. Politicians pontificate on the concept of the big tent, but pop commerce actually pitches it, finding a way to make square things hip and alternative things mainstream, selling work boots to hipsters and body piercings to insurance agents. You pays your money, you takes your titanium stud through the frenulum. In 1951, a man bought a pickup truck because he needed to load things up and move them. Things like bricks and bags of feed. Somewhere along the line trendsetters and marketers got involved, and now we buy pickups—big, horsepowered, overbuilt, wide-assed, *comfortable* pickups—so that we may stick our key in the ignition of an icon, fire up an image, and drive off in a cloud of connotations. I have no room to talk. I long to get my International running in part so I can drive down roads that no longer exist.

No pickup should endure the humiliation of being passed through a car wash. I was raised on *working* pickups. One pickup in particular—my father's 1971 Ford F-100— dominated my life from the time Dad purchased it when I was four years old until the day I left the farm for college. Saturdays, I shoveled it full with corn and oats. Dad took them to town to be ground, and when he returned from the mill, the bed was stuffed with feed bags twice the circumference of a tackling dummy. We lugged them one by one into the barn. Other times Dad returned with bags of barn lime, or a pallet of salt blocks, or bundles of baler twine. Once when we repoured the barn floor he came home loaded down with so many bags of cement that the frame was on the axles and the truck looked like a half-sprung lowrider. In the winter we loaded the truck with hay and drove through the sheep pasture, parceling alfalfa off the tailgate flake by flake. The following autumn, Dad mounted the side racks and set up a ramp, and we'd shoo the lambs aboard for their journey to the stockyards in St. Paul. My brothers and sisters and I used to spend hours slinging the truck bed full of firewood from the slab pile by the sawmill. Back at the house we'd reverse the process, unloading and stacking the whole works. I am prejudiced to the idea that you must work a truck to deserve a truck. I am prone to sneer at any truck with comfortable seats or, for that matter, a comfortable driver. I want to say, No, No, *No*. Pickup trucks should not be easy. Pickup trucks should be *hard*. This tendency is self-centered, unattractive, and, more to the point, irrelevant. I am beginning to think that once you hit forty, you spend the bulk of your time suppressing the urge to harangue everyone who comes within forty feet of your porch.

◊ ◊ ◊ ◊

In an effort to help me arrive at the age of eighteen alive and financially solvent, my parents quite wisely forbade me to buy my own car while I was in high school. Which meant I did a lot of dating in the F-100. By the time I got my license, the truck was entering its second decade of hard labor. The side panels were ragged with rust and flapped like buzzard wings. When you gained speed, they flared. The truck pulled drastically to the right. I'd hang off the left side of the steering wheel to keep it moving on a straight line. The transmission, originally three-on-the-tree, had been replaced at some point and converted to a stick shift accessed through a hole cut in the floor. No one is clear on why, but the mechanic put the new transmission in backward so that the gear selection pattern was reversed. You had to go far right and back for first gear and shift against your intuition. There was a gap in the floorboards beside the clutch through which you could gauge your speed based on the road blur. When it rained, my pant legs were mud-spittled. On the sharper turns, sheep ear tags and fencing staples shot across the dash. The brakes were inconsistent. Sometimes the pedal was as soft as squishing a plum. Other times the brakes caught so abruptly that empty vaccine bottles rocketed from beneath the seat and smacked you in the ankle bone.

Naturally, the windshield was cracked.

The heater was passable, but in the summer you'd rely on what a laughing bus driver once described to me as a "2–80" air conditioner: "Roll down two windows and go eighty miles an hour!" There were vents on either side of the cab at shin level, but to open them was to unleash a cyclone of alfalfa chaff and dehydrated horseflies.

Picture your date perched beside you on a summer's day, her lips glistening with Bubble Gum Lip Smackers and the cab charged with the scent of Gee, Your Hair Smells Terrific! shampoo. You're running fifty miles an hour down a gravel road when she grows overwarm and bends down to crack a vent. When she rares back, she appears to have emerged from a polluted wind tunnel. Her hair is frosted with feed dust and she's got pine needles stuck in her banana clip. Her lips are dotted like twin strips of flypaper, and there is a June bug in her braces.

You're young. You kiss her anyway.

I spent so much time dating in that old truck, I didn't know how to act in anything nicer. Once my grandfather lent me his Ford LTD. It was a beauty. Bloodred paint job with a white vinyl top and air-conditioning. Power steering, power brakes, and a fully automatic transmission. I was dating a farmer's daughter with the cutest button nose. I had coupons, so we got dressed up and went to Pizza Hut. After dinner I pulled out of the parking lot, merged into traffic, leaned back expansively, and draped my right arm across the back of the seat. The girl smiled up at me sweetly. She had grown to tolerate the farm truck, but as we picked up speed, I could see her luxuriating in the smoothness of the LTD. At which point, out of reflex and forgetting I was driving an automatic, I went for second gear, instinctively mashing what should have been the clutch but in the event was the power brake. I had my seat belt on. She did not. The image that endures is of her flailing elbows as she fought to unwedge her button nose from that pinch point where the windshield and dashboard meet.

I am a ridiculous pack rat, and it took me two hours of digging through boxes stored in the crawl space above the garage, but I found the Alpine Management letter. It was a kick to reread it. It is one thing to fall for a practical joke hook, line, and sinker. It is another thing to get hooked, landed, stuffed, and hung on the wall. After Eric's wife hung up the phone, all my anger evaporated, replaced by a sweeping feeling of relief and admiration. I was alone in the house, hand still on the phone, but I was grinning. A successful practical joke relies on art and craft, and this was a masterpiece. The bona fide parking ticket (originally obtained by

Eric when he got a parking ticket and the officer filled it out so lightly that the carbon didn't transfer) with its perfectly planted details, the tone of the letter, the carefully chosen nonexistent address, it all worked, but the genius touches were all those little shots to my ego (on the ticket under "VEHICLE COLOR," Eric had written "*RUST*"). I thought of myself huffing and puffing in the parking lot, speechifying at the windshield, making all those passes up and down Grant Street and running my trembling finger over the city map. The setup was brilliant, but in the end, the joke worked for one simple reason: it was predicated on the fact that my truck is *ugly* . . . and I love it so.

○ ○ ○ ○

Now February is gone and the truck still sits there. A monument to my dithering. There is a sort of informal open door at my parents' farmhouse every Sunday evening, and sometimes several of us kids end up there at once. Last weekend my sister Kathleen was there with her husband, Mark. I mentioned my desire to get the truck running and he said he would help. He and Kathleen have just had their first child, so Mark's going to be around the house more, I guess. Mark is a hot-rodding machinist and NASCAR fanatic. He proposed to my sister by faking a breakdown in a mud bog and requesting that she fetch a wrench. When she opened the toolbox there was nothing in there but a diamond ring.

Chapter 4

MARCH

FIVE DAYS INTO MARCH, and it is ten below zero. On my way out back to dump the compost I discover a plain cardboard box wedged between the doors. The box is about the size of a boutonniere carton. I recognize the logo on the shipping label. My garden seeds. I wonder how long they've been freezing here. Must have been a substitute UPS person. The regular UPS guy never leaves anything out back. A trace of snow has sifted in around the box and the crystals glitter in the sun. I imagine all the little seeds exploding into irrepressible green.

The dog chained to the house across the alley has spotted me and gone to barking. The dog is owned by a skinny man who keeps his mullet tamped down with a NASCAR cap. Thin as he is, the man moves with a stiff muscularity that implies hard luck and fistfights. The dog appears to have been raised on a diet of chain-link fence and burglar heels. Back when the weather was warmer, the man set out to build a kennel from steel hog panels, but he was halfway through pounding stakes when one of the women sharing the house with him stuck her head out the door and said something. He threw his hammer down and stomped inside and that was it. The panels still lie flat there beneath the snow. One day last fall some of the man's pals drove a white van up the alley and parked it beside the clothesline. The man met them with beers, and they popped the hood. By late afternoon, the yard was a scatter of parts and tools and every door on the van was open. The men were gone. When I walked my trash bags over to the village dumpsters across the street, I grinned at Matt, the village employee with whom I serve on the fire department, and nodded toward the van.

"Whaddya figure?"

He didn't miss a beat. "Two weeks, and it's a yard barn."

"Never leave the lot under its own power," I added.

Months now, and the van is still there. The doors are shut, but through the windows I spy Hefty bags and duct-taped boxes.

I could do without that dog, but I do not object to the hog panels or the van. I would be a jerk to do so, what with my International still lodged there out front—out *front*, mind you, not even hidden alley-side—of the garage like a prehistoric carbuncle. What that white van does is take the pressure off. When my clothes dryer died last winter, I wrassled it up the basement stairs and knee-tossed it out the back door. It landed off-kilter in the snow and stuck, like a dotless dice cube frozen mid-tumble, in a position reminiscent of the fifteen-feet-tall steel cube erected by the sculptor Tony Rosenthal on a traffic triangle in New York's East Village. I saw Rosenthal's sculpture once from the back of a cab. It is balanced on one point and can be spun on its axis. He put it up in the late 1960s and named it *The Alamo*. This is the sort of willful obscurantism that hinders the appreciation of modern art in the heartland. Apparently three-dimensional squares *en pointe* were trendy in the late 1960s, because a year later a cube designed by Isamu Noguchi was installed in a nearly identical position outside a high-rise a few blocks uptown on Broadway. Noguchi painted his cube red and named it *Red Cube*. I like Noguchi a little better for that. Riffing off Richard Serra's *Tilted Arc*, I took to calling my dryer *Tilted Lint*.

The dryer (or *installation*, if I may) remained in my backyard well into summer. When the snow melted, it settled to the ground. *Gotta get rid of that*, I'd think, every time I had to circumvent it on my way to water the garden or cover an ambulance call (the Serra reference was becoming more apt—*Tilted Arc* was removed after people got sick of detouring around it). I began entertaining fantasies in which I would preserve valuable landfill space by repurposing the dryer. It would make a fine industrial-strength compost turner. I'd trade out the belts for a chain drive and hook it up to my old Schwinn Varsity. I would schedule

organic spinning classes for my friends in the renewable energy crowd.
On a less vegetarian note, I also believed that given twenty minutes with
a cutting torch, a welder, and a bundle of rebar, the dryer could be converted to a monster hibachi with rotating spit. I'd hire out for weddings
and pig roasts. Alternatively, it might do for a deer blind. I'd have to strip
the guts and motor and drill out peep holes, but then I'd be good to go.
Just hide inside, and when a big buck wanders by, pop the lid and *rat-
a-tat-tat*. In this county the idea of camouflaging yourself inside a dryer
is not at all absurd, as the local forest creatures have grown blasé about
the presence of home appliances in the wild. Like overgrown cubist
toadstools, feral refrigerators and washing machines are generally found
clustered at the bottom of eroded gullies, or at the terminus of dead-end
roads and abandoned driveways. Trouble is, discarded white goods make
popular targets, and are often so ardently perforated as to appear to have
been caught in the crossfire at Panzer fantasy camp. Come November,
when the woods are filled with trigger-happy amateurs sporting blaze-
orange bomber caps, your perspicacious Leatherstocking does not take
shelter in a Maytag.

Ultimately, I sent the dryer away with the village junk man (the title
is unofficial, but the work is steady). You don't *schedule* the junk man as
such. He keeps his trailer parked at the implement store out on Highway M, so you call out there and they pass the message the next time
he drops by. If you're not home the first couple of times he stops, he
doesn't sweat it, because he knows your dryer isn't going anywhere. He
tows the trailer behind a teensy pickup truck, which magnifies the fact
that the trailer is roughly the size of a volleyball court. Fully loaded, it
has the appearance of a postapocalyptic Costco on wheels. Depending
on the nature of your junk and the going price of scrap iron, he'll dispose
of larger appliances in exchange for a modest donation. He also accepts
odd lots of steel, aluminum, and old wire. I don't know much about the
junk man except that he lives in the trailer court and his truck used
to have Missouri plates. He has an Appalachian drawl and a smoker's
cough, and he gets short of breath quick. I helped him fight the dryer
aboard the trailer, jockeying it back and forth until it fit between two
beat-up washing machines and a harvest gold oven. He was wheezing

pretty good when I handed him a few bucks, and before he eased them into his pocket, he took time to crease and fold each bill neatly, almost as if he was buying time to catch his wind. But when he pulled out of the driveway, he leaned out the window to grin and wave. Before he heads for the scrap yard he'll pop the shields off the washers and dryers and stoves and yank all the wiring in order to strip out the copper, which sells for a higher rate than the steel. He does this day after day after day. I see him running all the time. I can't imagine the grind of his week, or what he'd rather be doing, but every time that truck passes by he's got the hammer down, and he'll always return your wave. Lacking an ACT LOCALLY bumper sticker or hemp shorts, he nonetheless manages to do the right thing for Mother Earth.

○ ○ ○ ○

First chance I get, I take the seeds to my basement and set them to sprout. From November on, I look forward to the day I can flout the ice and snow by puttering under the lights of my gardening bench. It is all I can do not to jump the gun. I get so hungry for green, sometimes I plant things way too early. This year, I came home from watching the Super Bowl at a friend's house and scattered some pots with last year's leftover oregano and sweet marjoram. I have nursed trays of stunted lettuce on a windowsill in January just so I could pick a leaf and hold it on my tongue while observing the formation of snowdrifts.

But today I want to cheat the seasons in earnest, so I scratch a match across the concrete floor and ignite two of three burners on the portable LP heater (the third burner has a habit of howling like a demonic calliope) and position it at the back of my steel folding chair. Then I plug in a plastic boom box at the workbench. The boom box won't play CDs anymore because I karate-chopped the lid during an embarrassing spasm of rage triggered when it began skipping tracks, driving home the fact that I got exactly what I deserved for buying a $24.95 piece of outsourced superstore junk despite knowing full well as I stood there in the cavernous aisles of the High Church of Cheap Consumption that any money saved would one day be expended threefold on blood pressure medication and knuckle stitches. I own two of these chintzy electronic farces. Both have fractured stubs where the CD lid used to attach. As slow learners go, I am a real drooler.

Despite my Samsonite gorilla act, the radio receiver still functions. For basement puttering, I split my time pretty evenly between public radio and Moose Country 106.7. I like that Moose Country. They play the one-namers: Waylon and Willie. Buck and Merle. George and Tammy. Loretta. It is silly to say bad things about popular music, but for the record, Johnny Paycheck is to Kenny Chesney as corn whiskey is to wine coolers. This new stuff suffers from overgrooming. Even the redneckiest tunes ring tinny. One sometimes fears the lyrics of the latest busted-heart song were transposed from a marriage encounter handbook. It isn't that today's superstars aren't talented and hardworking. It's just that their way of doing things has passed me by. I look at the pretty cowboy on the Jumbotron and think, It is one thing to polish your craft, it is quite another to wax your abs. Recipe for the real deal: Combine two parts busted heart with one part busted knuckles, sprinkle with cheap trucker speed and crushed Valium to taste, and marinate in hard luck and leaky motor oil. Stir in Genesis and Revelation, add a dash of hope, and finish off while being forcibly evicted from a hotel bar. Hello, Tanya Tucker.

The local public radio station is on the opposite end of the dial. Literally, and however else you wish to parse it. The Venn diagram of listenership may come up a little short on overlap, but I'm happy to go on record as supporting both formats (it being only fair to point out that both have supported *me*). Near as I can tell, the commonest complaint about public radio is predicated on the inconvenience of encountering opinions in conflict with your own. That, and unctuous tone. Indeed, the NPR snootiness sometimes unhinges my own inner redneck, but in general I defend their usually steadfast refusal to dumb down. One does not ask Alistair Cooke to do the Chicken Dance. Any reservations I had about the format remain neutralized to this day by the fact that during the OJ Simpson trial, NPR was the only place to which I could tune at the top of the hour and expect a newscaster to lead off with *news*.

I will listen to NPR today because I have brewed a mug of green tea, and nothing says public radio like green tea. I dial the needle leftward until I pick up the unmistakably civilized tones of WHWC 88.3. I roll the tuner back and forth, easing the stereophonic sound to full swell. It shortly becomes clear that the host and guest are discussing the Rwandan genocides of 1994.

Pulling nested trays from a shelf beneath the bench, I unpeel them one by one and begin to parcel out the potting soil. After shaking the trays to settle the soil, I press the eraser end of a pencil into the center of each cell, creating a depression to receive the seed. Then I place the seed company shipping box on my lap and open it. The packets are in a uniform row. Fingertipping through them Rolodex-style, I pull out the ones I think will benefit from an early start. Leeks. Peppers. Tomatoes. Peel back the gummed flap, tip the seeds into my palm. Pinch them one by one and drop them in the pencil dimples. Top them off with another sprinkle of potting soil and a pat, then move on to the next seed packet. I take my time. There is no clock in the basement, but the top of the hour is marked by a chirp tone that triggers local station identifications, followed by the familiar, *"From NPR news in Washington . . ."*

The familiar voice will always have a familiar name. *"From NPR news in Washington, I'm Karl Kasell . . . Anne Garrels . . . Linda Gradstein . . . Brian Naylor . . . David Welna . . . Craig Windham."* The tone is always unhurried, and eminently civil. Some of the names are poetry in themselves: *"Korva Coleman . . . Don Gonyea . . . Mara Liasson . . . Sylvia Poggioli . . ."* Those last two, if I had first seen them on paper, I would have mangled the pronunciation. Having heard them again and again in the newscasts, I can recite them flawlessly. Same with Frank Stasio and John Ydstie. I get a special kick from Corey Flintoff and the specific care with which he pronounces his own last name, floating down into those twinned final *f*'s as gentle as parachute silk settling to the ground, deflating with just enough force to generate the velvet fricative. Flint*ohhfff*. NPR should consider marketing a CD featuring Corey Flintoff repeating the phrase *alfalfa foofaraw*. Over and over, on a loop. The effect would be similar to those sound conditioner contraptions that lull you to sleep with the sound of electronically generated surf. When Lakshmi Singh does the news, I find myself saying her name just to hear it. *Lakshmi Singh.* She'll be leading off the newscast and she'll say, *"For NPR News in Washington, I'm . . ."* and I'll jump right in rhythm and say,

"Lakshmi Singh!" The part where the consonants *kshm* mesh and take us from the vowels *a* and up to *i* is luscious. She will sign off, and five minutes later I am still reciting the mantra: *Lakshmi Singh . . . Lakshmi Singh . . . Lakshmi Singh . . .*

Ah, but if NPR must be reduced to one voice, let it be Shay Stevens. All others are chattering children by comparison. Shay's voice is the personification of strength and reassurance. Warmth, ease, and a dusting of rasp. You imagine the enfolding motherly bosom, sensual but steady as she goes. I reject your big booming boys and nominate Shay Stevens for the Voice of God.

○ ○ ○ ○

This basement of mine will never be one of those wood-paneled airhockey-and-Ping-Pong-table romper room basements. The ceiling is low and dungeony. The joists are flossed with cobwebs. Much of the year the floor is damp, although it dries up nicely in winter and is dry today. The house was built in the 1930s. Old-timers tell me contractors upped the ratio of sand to cement in those days to save money. I don't know about that. Could be, because the concrete in the walls is cappuccino brown. Despite this, the house is solid and square. Down here beneath the frost line, I feel cocooned and safe. Enjoying, as poet Bruce Taylor wrote in his *Pity the World* gardening poems, ". . . some privileged ignorance of/the hungrier facts of life." I am grateful for my safe little spot, but freshly reminded of the Rwandan horrors, I am humbled by the fact that my gratitude alleviates no one's misery.

Last year, I bought Nancy Bubel's *The New Seed-Starters Handbook*. Trying to better myself, as usual. I read the preface earnestly. I felt a swelling sense of purpose. But then I began to skim, and then skip ahead, and naturally, I became overwhelmed. The old cookbook overload kicking in. It was frustrating, but it also made me appreciate the commitment of years and time required to become a true gardener. In the introduction, Nancy Bubel says she began gardening in 1957, when she killed a batch of radishes in a window box. It was nice of her to include that. I am learning not to overstudy, but rather to be satisfied above all by the process. Learn what I learn, on the fly. I try to choose wisely, plant things on the basis of the growing cycle and need for a head start, but pretty quickly I am throwing in favorites just because I miss them. Here in March I hunger for the stem crack of cilantro plucked in the morning sun and the thrill of discovering a deep green zucchini squash lying boa-belly fat in the grass, and so I unseal their respective envelopes and tap out a palmful of each. I do this fully recognizing that cilantro planted directly in the ground will catch and pass the wan, leggy stuff I'll sprout in these trays, and that starting your zucchini early is like starting your dandelions early. The hunger here is not so much for the food as it is for the sight of the sprout. And so I spend a few quiet hours in the subterranean sanctity of my ratty basement, tamping hard seeds into cushiony cubes of humus, putting in motion the predictable miracle of germination. A miracle available even to a klutz like me.

In Rwanda, 800,000 died. Most of the killing was done with machetes, axes, and hoes. One by one. Face-to-face. Neighbor on neighbor. Here I am with my seed packets. You can bow to the six directions of the cosmos full-time until you blow your back out and never reconcile the capricious distance between dumb luck and utter horror. In the preface to *The New Seed-Starters Handbook*, the poet-farmer Wendell Berry says that growing your own food is a sacrament. A visible form of an invisible grace. It is certainly an act of faith. When you tuck that seed in the dirt, you are drawing on the past to bank on the future. In another of his gardening poems, Bruce Taylor writes that planting serves "to bring us to our knees/to bring us back to quiet . . ." No matter the speed and uncertainty of the approaching future, we love to put our hands in the dirt, "where there's little/choice but to begin/with the intensive/care of the present . . ."

I take the trays upstairs, to a southern-facing window, and place them in the light.

❍ ❍ ❍ ❍

Here lately I weep more easily. There is a sea change happening in my heart. Nothing too dramatic. I rarely blubber or sob, but I tend to well up on short notice and in odd—sometimes ridiculous—context. I get misty at the sight of an elderly woman smoothing an old man's hair, or a suburban tot picking out Halloween pumpkins while clinging to the arthritic finger of her grandfather, a gnarled farmer. Looking for a laugh and lured by the casting of Rowan Atkinson as a bumbling priest, I borrowed a friend's copy of *Four Weddings and a Funeral* and wound up so unexpectedly afflicted with sniffles over John Hannah's recitation of W. H. Auden's "Funeral Blues" that I watched it twice more just to see if it had the same effect. It did. In another baffling moment, I recently became teary while reminiscing with a goose hunter in a cow pasture. And not so long ago I was in a snazzy hotel room far from home watching a scene from a television documentary in which a bail bondsman has taken time out from collaring thugs to visit his newborn grandchild. The hospital room was jammed with family, all of them—right down to the new mother—looking like they had jumped bail a time or two themselves. But there was something in their beaming faces and the burly man's eyes as he hugged the infant to his neck that broke me down. Tears were slip-sliding down my cheeks even as my inner Norwegian said, *Get it together, son, you are weeping over a documentary on the bail bond industry*. I was so startled by my weakened state that I called room service and ordered a plate of raw Kobe beef slices to restore iron to my blood. The beef arrived garnished with shavings of ginger, which cleared my head, although the piney notes, as they always do—especially when I am alone away from home—reminded me of fresh sawdust in the sun, and I was tempted to resume weeping, this time for the dear departed sawmills of·my youth.

The radio show about Rwanda sets a sadness in me that will recur for days, compounded by my awareness that such moping is at best impotent and at worst cosmically insulting to those who suffer whether you mope or not. Whether gardening safely in your basement or staring point-blank at the rotting corpses, you simply cannot relate. And yet it is the very voluminous evidence of the horrible things we willingly visit upon each other—what author Philip Gourevitch has described as *uncircumscribable* horrors—that invests a willing act of kindness, the tiniest touch or gentle word to a friend or stranger, with energy powerful enough to reverberate around the universe. And so when I see acts rooted in gentleness and purity, the tears rise.

In part I suppose this is all driven by chemical changes associated with advancing age, but I think also the loosening of tear ducts is tied to everything good ever squandered—the headlong accumulation of which instances you cannot help but note if you live attentively. I can tell you the tears are not bitter. They feel something like relief, and as such, I am lately forming an idea that they are triggered by glimpses into some fourth dimension—perhaps string theory is involved—when we sense how all time and experience is joined. If we live heartily enough to take on some scuffs and disappointment, we develop a yearning for the soulful moment. Perpetually poised at the vanishing point of a yawning infinitude, we come to see that our only lasting powers are love and hate. I tear up over the bounty hunter because as he holds that baby to his cheek, I believe he senses the inherent fragility of the moment, and how quickly it may shatter. The crack of a rifle in Memphis and a million dreams die. My friend the goose hunter may be understandably nervous about my glistening eyes, but the dampness on my lashes tells me I am alive.

○ ○ ○ ○

I am happy to live in a place where I can chuck a washing machine out my back door and no one judges my behavior unusual. Having said that, I recognize the limits. Shortly after I moved to this village, I was upstairs writing one afternoon when a steady rhythm of thuds gradually intruded on my conscious. Moving to the back of the house, I peeked through the blinds and saw two teenagers, each armed with a sledgehammer, pounding the bejeebers out of a junk car in the adjacent yard. To compose the approximate image, visualize a pair of manic first-chair kettle drummers slamming madly through a speed-metal update of "In-A-Gadda-Da-Vida." The bulk of their blows were directed against the hood, trunk, doors, and roof, so the clamor was largely metallic, but now and again they'd strike a headlight, and the tinkle of glass came through like a grace note. When they drove the head of the sledge through a window, the safety glass gave out a squeaky crunch and collapsed into sparkling honeycomb. The boys hammered until every flat surface had been thrashed to a rumple. Then they set aside their tools, and, as men are given to do, retreated three steps to gaze upon all they had wrought.

As long as there's a shot your car might one day spin a wheel, the village board will generally give you dispensation to leave it parked in perpetuity. But when you pulverize your Pontiac to the point that it appears to have been tumble-dried in a rock crusher, you are sending a specific message, and that message is: *this vehicle has been rendered irretrievably out of service*. Someone complained, and the village board commanded that the vehicle be removed.

Occasionally this happens—someone will show up at a board meeting and ask that a patch of weeds be mowed or a car hauled off. There are ordinances, and I understand, to a point. Neatness keeps the property values up, and based on the number of scrawny cats emerging from the three-foot foxtail across the alley to dump a load in my cilantro, you could probably advance the argument on public health grounds. Still, I am leery of enforced neatness. We're seeing more and more of it around these parts. Zoning, covenants, "smart growth," and so on. We strive to preserve the countryside. Minimize the impact of big boxes and sprawl. But social engineering in the cause of perfection and the tax base has its casualties. You can't build a simple shack without a series of visits by some bureaucrat waving a sheaf of permits and a clipboard. People move here from the city and put up their dream home on twenty acres and don't want to gaze upon tin siding and caved-in Plymouths. I get it. But neither do I want someone checking my quack grass with a tape measure. To my eye (and I freely cop to a festering case of latent hickitude), that trailer house tucked against a row of Norway pines is far less objectionable than some tony monstrosity dwarfing the ol' fishin' hole. Gentrification is not always a matter of Starbucks. When my brother four miles north of town is informed that in certain circumstances all of his neatly stacked lumber must be a minimum of twelve inches off the ground, I can't help thinking some people have too much time on their hands and our tax dollars would be better spent on the local kindergarten teacher. My brother, whom I'm sure appreciates that I "can't help thinking," tends in nearly all cases to adopt firmer courses of action, and is standing for election to the town board. We are not of one mind on all issues, but I solidly admire his willingness to take the abuse.

He and I are both complicit. Him with his bulldozer, with which he carves driveways for the new arrivals, me with my little ten-acre patch outside of town, which I sold in a trice when I needed the money and found out what it would bring. And if the time comes to put my house up for sale, I will be seeking the highest bidder, which (based on what I paid for the place) will raise the property values accordingly, continuing to prove the point that no matter who's shinnying up the trunk—land-hungry developers or preservationists intent on legislating the position of every pine needle—it's the poor folks who get pushed from the tree. In the meantime I ponder the advantage of keeping one's place in a state of rattiness capable of evincing sympathy from the assessor. He sees a rusty International and a marooned dryer, I see a pair of tax deductions. I shall reserve the money saved for a quick spruce-up when it's time to sell to a buyer looking for a decent place with affordable taxes. Arrange a situation *à la* Chevy Chase in *Funny Farm* in which the neighbors pitch in by nailing their siding back on and disguising the pile of rusty carbure-tors and two-legged Weber grilles with a hand-stitched Amish quilt and a smattering of heirloom squash. Just long enough so you can close the sale with some nice young couple escaping the big city.

Time passed, and once again I was drawn from the keyboard by an apoc-alyptic clamor. First, an engine, revving to the point of warping the valve covers. Then the sound of spinning wheels, general acceleration, and the faint rattle of chain links. Finally, the whole buildup terminating in a horrifically conclusive *ha-WHUMP!* I rose from my desk and assumed my customary position at the upstairs window.

The family van had been brought around to the backyard, where a crowd had gathered around the annihilated Pontiac. The van was one of those customized jobs with flare fenders, tinted picture windows, and a silver luggage rack. It had long ago gone to rust and sag, but remained the pride of the family fleet. The van was backed into a position perpen-dicular to the driver's side of the car. A logging chain was hooked into the frame of the car somewhere under the passenger side, drawn up and wrapped around the passenger side door, pulled across the rooftop, and then angled down to the rear bumper of the van where it was secured to the trailer hitch. I was trying to make sense of the arrangement when the van shot forward.

Ha-WHUMP!

The pummeled car lurched six feet sideways and the van stalled. The driver restarted the van, backed up, and roared forward again. *Ha-WHUMP!* The process repeated itself over and over. They yanked that car around the yard for ten minutes. And somewhere in there, it hit me: *They are trying to flip the car.* By passing the chain from the bottom of the car and over the top of the car and then using the van to jerk the slack, they theorized that enough lateral force would be generated to spin the car on its long axis and flip it wheels-up.

And on maybe the thirty-seventh try, the whole ridiculous project panned out. In a perfect convergence of bounce, torque, and frame-snagging topography, the *ha-WHUMP!* was followed by a beat of silence during which the car teetered on the driver's side, then a much softer *whump* as, pushed past the tipping point by several alacritous bystanders, it flopped on its flattened roof. One wheel spun slowly. After a brief round of congratulatory whooping and fist-pumping, the boys moved in with tire irons and started removing lug nuts, and then I understood: they had hatched this great endeavor in order to salvage the four tires, which, despite all the violence, still held their air, if not their tread. For lack of a jack, it had been deemed simpler to flip the car. Five minutes later the tires were leaning against the garage and the automobile was being trailered to the salvage yard.

What you had here was a full-on testament to the blue-collar work ethic, Yankee ingenuity, the fundamental beauty of main force violence, and the potential upsides of beer and unemployment. I felt the urge to salute. Recently an outsider questioned my neighbor Trygve about the local propensity for cockeyed perseverance in the face of absurdity. "We up in New Auburn," replied Trygve, "walk a more difficult path."

○ ○ ○ ○

We've hit a warm stretch. The fourth Sunday in March, and temperatures are headed for the mid-60s. Nearly all of the snow is gone, and when I open the kitchen window, the air that rolls through the screen

is soft with the watery scent of melt. I step out the back door to feel the
sun on my skin. For the first time in months, the earth gives a little under
the sole of my shoe. I know this is not spring. It is the first of many false
starts. But the smell of the warm air and the sound of the soil loosening—that barely audible trickle and drip—taps a pinprick of melancholy
that blooms from my heart in a sweet, radiant bleed. The sensation is
vestigial of adolescence, when the first whiff of emergent earth triggered
fits of yearning and longing. Not lust, but rather the purest sort of want, a
nostalgic intuition that somewhere a woman waited beside a stone fence
beneath a wide sky, pining for me to stride from among the pine trees
and take her in my strong arms and my true heart. There would be a
popple tree thicket and a blanket, within and upon which we would recline with earnest intentions, and from somewhere in the bushes would
come the sound of Neil Diamond singing "Coldwater Morning," "And
the Grass Won't Pay No Mind," and "Girl You'll Be a Woman Soon," in
medley. That would be love, and that would be it. The scene would neither conclude nor progress, leaving me instead to hover timelessly in a
stasis of poignant bliss, chastely cradling my anonymous desideratum.

I trace the Neil Diamond thing to my aunt Meg. I recall playing air
guitar to her vinyl copy of *Hot August Night* during a family get-together
when I was very young and girls were yet a nonfactor. When the hor-
mones hit, I borrowed Neil's albums from the tiny library in Chetek,
Wisconsin, where I believe the record will also show I checked out every
Louis L'Amour, Zane Grey, and Max Brand Western in stock. Cowboy
books and Neil Diamond: hello, solitary man. These twin influences
shaped my concept of romantic comportment in ways that lingered into
my mid-thirties. Meaning, I owe a smattering of perfectly classy women
signed letters of apology. And yet I still pull out the Neil albums sometimes, if only to feel again what I once imagined love might be. "*Shiloh,
when I was young . . .*"

The Best of Bread will also do the trick.

When the weather warms, the village stirs. Cars passing up and down
Main Street run a little looser, their drivers uncurled from the deep-
winter heat-conserving hunch and more prone to raise one hand from

the wheel and wave. Last week the tires gave off a muffled crunch and squeak as they rolled over the snowpack. Today they roll on naked asphalt, grinding the residual sanding salt between the rubber and the road with a gritty crackle that echoes from the vinyl siding of the surrounding houses. I can hear children calling, radios playing, doors slamming. The *ska-winch, ska-winch* of trampoline springs. There is something lonely in the distant sounds, and now I am ruminating on the failures of affection and assessing the damage done, and my Neil Diamond reverie is derailing. The accumulated dings and cock-ups of love—once "breaking up" advances beyond merely switching school bus seats and results in real hurt—tend to leach the sweetness from melancholy. Two blocks over on Pine Street, the morning worship service at Bethel Lutheran is commencing. The tone of the bells is mellow in the warmth. My back-alley neighbor steps out the back door and snaps a leash on his ferocious dog. They set out down the alley, the dog lunging, the man leaning backward, a decorticate stutter-stepping water-skier. Still, he manages his aura of detached toughness. I can see words across the chest of his T-shirt. He gets closer, and I can read them: 100 PERCENT WHUP-ASS.

The resurrection of the International has officially commenced. My brother John used to work for the local implement dealer, and the dealer has allowed us to borrow his equipment truck. It has a tilt bed and a winch. John backs into my driveway, then, using controls mounted on the truck's frame, runs the bed back and down until its beveled tail is wedged behind the rear tires of the International like a stunt ramp. Unspooling cable from the winch, he slings the hook around the frame. Then, with me in the International to steer, he rewinds the cable, pulling the pickup slowly backward up the ramp. I can't help but grin like a kid. For the first time in years, I am sitting inside the International while it moves. When all four wheels are on the wooden deck, John retracts and lowers the bed. We secure it with chains and boomers and—after pausing for purposes of a good story to measure the depth of the depressions in the asphalt where the truck has stood for all these years (officially: three inches, per the Stanley tape)—are on our way. Up in the cab of the implement truck, John's dog Leroy rides shotgun, gazing loll-tongued and happily out the window from his elevated position in the passenger seat. I follow in my car.

It is something else to see that old truck up there, moving down the road high in the air and backward. Exposed in that manner, it looks surprisingly frail. The sky is pale blue and clear, the sun's heat today just adequate to ease a small seep of melt from the one shrinking blob of snow remaining on the hood after all the recent warm weather. The trip to my brother-in-law's shop is just over twenty miles, and John takes it easy, rarely running over forty. I settle in for the ride, turn on the radio, and punch up Moose Country. When you are rolling down the road behind your brother headed for a truck revival, you want to hear some chicken pickin'. *Whoop-whoop-whoop*, said the late Waylon Jennings, and, *boogity-boogity*.

Mark is waiting outside his shop when we arrive. John backs in, raises the bed, and decants the truck. I ride in the cab again, giving Mark a thumbs-up and steering as the unspooling cable lowers me to the ground. When the truck is on the level, Mark and I brace our butts against the rear bumper and push it the rest of the way across the concrete apron and into the shop. Then we troop into the house. My sister Kathleen has made a batch of spaghetti. After the crisp air it is good to come in to the pasta steam. Mark and Kathleen's house is built into a hillside, and from the dining room table we can see for miles across the simply and perfectly named Blue Hills of northern Wisconsin. We talk about the truck, but we also talk about our cousin Sukey, a helicopter pilot in the U.S. Marines. She sends us e-mails from somewhere in Africa. We remember her as a blond tyke at Fourth of July picnics. Her husband, Steve—also a helicopter pilot in the Marines— is facing deployment to Afghanistan. Mark and Kathleen are adjusting to their newborn boy. I call him Sidrock, which is not his given name, but such are the prerogatives of uncles. While the grown-ups talk about loved ones at war, Sidrock is goobering in his infant seat.

Mark and I plan to start tearing down the International next weekend. The idea is to get it back up and running by November, in time for deer-hunting season. I have this desire to bring my deer—a year's provision of meat—out of the woods in the back of the truck. On the way home,

I am following John down a dip between two hills when he slows on the upside and gradually stops. Out of gas. John has already burned half a day helping me, but when he climbs down from the cab he is shaking his head and grinning. "I've got a couple cans back home," he says. Rather than try to explain to me where to find them, he and Leroy take my car, and I sit on the flatbed facing backward, ready to warn off anyone who comes over the hill too fast. We used to follow this road to church when I was a child. Back then this truck was new, delivering shiny tractors to every corner of the county. Now the farms are gone and the only tractors selling are smallish vintage models, repainted so they'll catch the eye of some out-of-towner when they pass by on the way to the lake property or hunting cabin. I place my palm on the worn wooden deck and it transmits a sweet ache. This is getting ridiculous. In the seventeenth century, nostalgia was considered to be a diagnosable mania. They'd be slipping Prozac in my porridge. The sun is warm but the wind is cold, and I'm glad when John returns and we can go home.

✿ ✿ ✿ ✿

The snow will fall again before the frost is out, but we are gaining momentum for that time of year when everything reappears. The ugly stuff first, as it happens: neglected lawnmowers, weed-wrapped tires, waterlogged copies of the *Early Bird Shopper*. But most days there is a flavor to the air that suggests greenery will triumph.

Last thing before I turn off the lights and head for bed, I inspect the sprouting tray. Each cell cradles a cube of peat, and at the center of each cube is a seed, quietly pursuing the sun. Thomas Moore said tears are a luxury only to the happy. When I am forced to cast my eyes beyond my own navel, I realize that a dip in sweet melancholy is every bit as indulgent as a bubble bath. Be joyful, says Wendell Berry, though you have considered all the facts.

Chapter 5

APRIL

I HAVE A DATE. With a real woman.

I have scheduled a haircut.

I was born with hair trouble. *"Looks mostly like Daddy,"* wrote Mom in my baby book. *"Has his front cowlick."* In a photo taken at three and a half months, I am sitting naked in the bathroom sink, my eyes are wide and my eyebrows are raised. I look like a startled little Buddha. The expression amplifies the cowlick, which flares in a backward arc over the right-hand side of my brow. In my second-grade school picture, I am sporting a buzz cut. The cowlick bristles like a crockery brush. In third grade, I grew out my bangs, and the cowlick popped through like crabgrass. I fought that thing six ways to Sunday. Plastered it flat, slept in a cap, hair-sprayed it, combed it silly. Nothing worked. I pestered my mother to the point that she took to strapping it down at bedtime with pink beauty tape—the kind Ladies of a Certain Age use to secure their rollers and pin curls. I'd sleep in the tape, peering hopefully into the mirror each morning as I peeled it free, then—*ptoing!*—it busted out like sprung watch works. I think of my father, quiet farmer that he was, slogging upstairs to bid his firstborn son good night with Bible verses and a kiss on the brow, only to find that brow swathed in pink hair tape. I wonder that he didn't just grab the phone and book me for *Scared Straight*.

In seventh grade I began to care how I looked, which is a shame. My first attempt at hairstyling was a modified Monkees mop worn in a left-to-right swirl. The swirl overcame the cowlick by flopping it sideways, a happy effect that lasted only until I nodded or the wind changed. Then in the 1970s, Shaun Cassidy parted his hair in the middle and launched the golden age of feathering. The nation and I went for it whole hog. For Christmas I requested a blow-dryer, and in my eighth-grade school portrait, I am no longer fighting the cowlick. Instead, exercising a form of follicular judo, I have turned the cowlick's own momentum against it, combing and blowing it straight back. This in combination with a velour shirt gives me the appearance of a youthful televangelist emerging from an explosion at the hair-spray factory.

This was also the era of the white man's afro, but my parents wouldn't allow it. This is theoretically too bad, because a perm was the one intervention that would have put that cowlick in a full nelson, but in reality they spared me one more set of bad hair pictures. (Or worse—while leaning over to notate his lab book during chemistry class, my friend Marco Bucklinski stuck his faux 'fro in the Bunsen burner. I looked up at the sizzle and snap and found him playing his head like the bongos.)

Eventually I backed the blow-dryer off HIGH and let my hair hang a little more lank. To the left of the part, it feathered back à la Shaun Cassidy. To the right, the cowlick rose and fell with a swoop. I took this to be mysteriously moppish in a *Sweet Baby James* sort of way and got in the habit of fine-tuning it with a sweep of my right hand, an idiosyncrasy that quickly hardwired itself into a permanent tic that persisted through my late thirties. By my senior year in 1983, I was leading the football team in sacks, but was forever fussing with my hair. My teammates voted me Most Valuable Lineman. My classmates voted me Biggest Primper.

During the college years, my hairstyles evolved under three central influences: Pre-Mellencamp John Cougar, Bono Vox circa *The Unforgettable Fire,* and once—when I teased and spray-painted my hair pink in order to establish street cred with the ditchweed-dealing ruffians at the roller rink where I was employed as skate guard and roller-skating Snoopy—the Great Hair Metal Scare of 1986, specifically as personified by the bands Poison and Cinderella. Ultimately, Bono's influence was most pervasive, leading me to scavenge Eau Claire County's lone mall for boots like the ones he wore in the *Pride (In the Name of Love)* video. The closest match I could make was a floppy-ankled pair from an all-women's shoe store. I take a ladies' size 10, as it turns out. I tucked my parachute pants in and wore the boots with an air of meaty goofball angst. It has only recently occurred to me that technically, wearing those boots counts as cross-dressing. Mistakes were made.

Your 1980s man had a multitude of hair options, and in May 1987, I achieved critical mass, graduating from the University of Wisconsin at Eau Claire School of Nursing with mousse spikes on top, a mullet in back, and a moustache up front. The bad hair trifecta. I gave the commencement address wearing a crumpled white linen suit and a pastel blue tie as wide as a pencil, then drove off to the future in my rattletrap pickup truck. The hood lettering said International, but my hair said *hot red Fiero*.

○ ○ ○ ○

Finally, *finally*, we are going to get started on the truck. It's a Saturday evening, already dark. Temperatures have dropped back to the low teens, and when I get out of the car in front of Mark's shop, I catch the scent of wood smoke. Mark has stoked a fire in the cast-iron stove. When I step through the steel security door, the warmth folds around my face and dissolves the stiffness from my cheeks. Across the shop, the flames pulse and waver behind the isinglass.

Technically, the shop is a *garage,* based on the fact that it is a freestanding structure equipped with a pair of retractable overhead doors and two bays designed to hold one vehicle each, but the vehicles are all outside beneath the oak trees. That battle was lost a long time ago, probably about the time someone wired the place with 220 and Mark hauled in the parts washer. The northwest corner of the building is dominated by an L-shaped wooden workbench and shelves built against the concrete block wall. The workbench is stained and splattered and generally dinged in a manner your high-end antique dealers will classify as "distressed." The wood is different shades of used motor oil, with a fresh white scar here and there where someone set the sidewinder grinder down while it was still spinning. In addition to a greasy phone book, the working surface of the bench is currently occupied by one empty coffee mug, an empty can of Coke, an empty can of Busch beer, one container of all-purpose glue and one container of all-purpose solvent (always leave yourself a way out), a tape measure, a scatter of wrenches, and, tucked beside the toolbox, a half carton of chocolate-covered Whoppers. Also in evidence on a clear spot: a pen, and a spiral notebook open to a thumb-smudged page of notes, numbers, and obvious figuring.

There is another tape measure on the floor.

A large rectangle of pegboard is screwed to the wall above the right-hand arm of the bench, and this is studded with hooks from which dangle assorted belts and pulleys, rolls of plastic line for the weed whacker (the weed whacker itself is hanging over there against the other wall), coil springs, bungee ties, a stapler, a miniature carpenter's level, grinding wheels, and several paintbrushes. Some of the items—plastic pushpins, a gasket kit, zip ties—hang in the same plastic packs in which they were displayed at the store. The rest of the space surrounding the bench—above and beneath, from floor to ceiling—is taken up by catchall wooden shelves. Beneath the bench you find heavier oddments: a splined shaft, a motorcycle battery, an electric motor, and various cast-iron thinga-mabobs studded with knobs and plumbed with gauges. One set of upper-tier shelves is strictly devoted to cans of spray paint; elsewhere you see spools of solder wire, bits of copper tubing, a car radio trailing wires, a hitch receiver, a pair of brand-new taillights, various flashlights, an anti-freeze tester, a broken watch, and that sacrament of the shop, a spray bottle of WD-40. Three emergency road flares are balanced on a streaky can of varnish, which is next to a used automotive coil, a hitch pin, and a pair of car speakers. A small gasoline engine sits on the shelf at an angle, as if it is edging toward jumping. There are also several parts organiz-ers with their small drawers arranged in rows and columns and filled with nuts and bolts and washers and rivets and cotter keys and whatever else fits, and a couple of shelves hold books: *Student's Shop Reference Handbook, Automotive Engines Maintenance and Repair, Machinery's Handbook Seventeenth Edition, Motorcycle Basics*, a Chilton manual, and several parts and accessories catalogs. Every flat surface is in service.

On the sill beneath the glass block window I can see a box of Band-Aids, a sanding pad, a pack of baler belt fasteners, some hose clamps, and a set of fluorescing shotgun sights.

Most of the tools are confined to a space beneath the heavy iron metalworking bench or stowed in the tall red toolbox wheeled up against one wall, although you do see a monkey wrench or plastic mallet lying here and there. There is a drill press on its own stand, a counter-mounted grinder, and a chop saw. A modest collection of steel stock leans against the wall beside the furnace ash can, and there is a V-8 engine bolted to a wheeled stand over beside the hydraulic floor jack. A stepladder leans against the south wall, where a row of hooks holds extension cords, trouble lights, logging chains, a cant hook, a buck saw, a snow shovel, and a dusty motorcycle helmet. Other hooks spotted variously around the walls dangle a lawn sprinkler, two begrimed tennis rackets, a vintage Volkswagen hood ornament, and the windshield from a purple Toyota Land Cruiser. Up against the west wall beside the furnace a rack of steel shelving is crammed floor to ceiling: I can see a rusty pair of shocks, an air rifle, a Coleman lantern, an ancient in-window air conditioner, a pair of paint respirators, a galvanized Miller High Life cooler, a radio-controlled truck, ice fishing equipment, an ax, a rainsuit still in the package, and a drive shaft.

If your experience with shops is limited, I should make the point that Mark keeps a clean work area. Beyond the two empty aluminum cans, the one thing you don't see is trash. The floor is swept, and anything that isn't useful or potentially useful is stuffed in a blue plastic barrel over by the door. The thing that strikes me as I look around is, you could run the revolution from this place. I would add that over in the southwest corner next to the chain saw is a refrigerator stocked with beer, and just above the parts washer, a miniature disco ball. After the revolution, you could party.

If there was any doubt that Mark is the right man to help me with the truck (he has already restored or upgraded several vehicles, including the purple Toyota Land Cruiser), they would be erased by the machine currently taking up space in the bay opposite my truck: a partially constructed sawmill, which he is building from the ground up using parts from a turkey slaughtering conveyor, an automatic car wash, and a motorcycle engine. The rest of the machine he is fabricating with his own hands. Mark is a machinist by training and trade, and you can see it in his work. The sawmill does not have the appearance of a homemade contraption. Because the raw steel stock of the frame and carriage is still unpainted, the underlying craftsmanship is open to inspection. The welds are neat and the angles are precise. Inside a tangle of hydraulic lines are gears and pulleys and rollers—many of which Mark machined himself—the placement of each requiring consideration of the placement of the others. The sawmill looks at once intricate and bomb-proof.

○ ○ ○ ○

Mark is about my height. Wears his hair in a self-administered no-nonsense buzz cut. Tends to run a goatee with attendant stubble. A scar just off the crown of his scalp indicates that at some point he took a pretty good shot to the head. He probably goes ten to twenty pounds less than me depending on how hard I've been hitting the doughnuts. He's lean and a little bowlegged. In the summer he wears knee-length shorts and leather work boots, and this accentuates his bow legs. He carries himself with the wariness common to men who express themselves on the factory floor more than the dance floor. In the presence of strangers, he will be closemouthed and tentative. You might interpret this as deference, but as any out-of-towner on the losing end of a tavern brawl can tell you, that'd be a mistake.

There is one anomaly in his appearance. Mark has classy eyeglasses. Nothing fancy, just wire frames, but tastefully cut. Say what you will about the common-denominator crudity of American mass culture, these days you see tasteful eyewear everywhere, from the Big Apple to Big A Automotive. It took a while. It has something to do with the fact that you can get a Michael Graves two-slice toaster at Target and a latte at the Gas-N-Go, but I also credit the Germans. I backpacked around Europe in the summer of 1989, and it seemed to me that of the ten or so countries I passed through, the Germans above all had a noticeable penchant for arty spectacles. I refer to it—especially in the case of superstar architects, trust-fund bohemians, and movie stars trying to cultivate an intellectual air—as *considered* eyewear. Arch, slim lines. A certain studied geekiness perfect for perusing the coffee table version of *Design and Form*. And lest I come off as catty, I plead that I came of age in the plastic-rimmed bug-eyed 1980s and this particular esthetic refinement couldn't happen soon enough. It's just that it has taken me awhile to adjust. You expect to see these glasses on, as a recent *Wikipedia* entry has it, what Peter Drucker called "knowledge workers," but not so much on a guy who builds his own sawmill and watches NASCAR races while drinking cans of Busch beer from the comfort of a recliner upholstered in Mossy Oak Break-Up camouflage.

The glasses may be refined, but Mark's gaze is unalloyed. It is clear-eyed and direct. When he looks at you, he is sizing you up. He eyeballs you the same way he eyeballs a length of channel iron, gauging where he should make the cut. Sometimes when he just stands there dangling a three-quarter-inch Craftsman, I can imagine him bringing it down on my skull. This I keep to myself.

I simply cannot resurrect this truck without Mark. I don't see things right. Never have. Even when I was working on my dad's farm or as a ranch hand in Wyoming, most of the mechanical skills I developed were dependent on simplicity and repetition. I learned to reconstruct the header on my hay swather because it was forever beating itself to bits, but even then I was dealing mostly with the most fundamental sorts of problems—spun bearings, cracked steel, busted sickle teeth. All repairable with a wrench, a hammer, or a slather of bubblegum welding. Faced with an occluded grease zerk, I could replace it, even tune up the threads with a tap wrench. But if the engine started knocking, or wouldn't start for any reason other than a dead battery, my only answer was to go begging for help.

I love the work. Love to get in there. Love grease on my hands, bark marks on my knuckles, bits of stuff in my hair. Love to see the whorls of my skin outlined in black, a topographical map in the palm of my hand. I like the feeling of lying on my back beneath the chassis trying to reach a rusted nut with the heat of the trouble light in my ear, squinting and holding my breath against the burnt dust and grease that rises off the naked bulb in a twist of white smoke. I know how to *work* like a mechanic. I just don't know what to *do*. At best, I am a good helper. A hander of tools.

So I stand beside Mark, and we study the truck. It sits in the bay opposite the sawmill. We don't tear right into it. We stand back, hands in pockets. This is the time you knock your cap back and study things out. The truck looks a hulk. I happen to know it can reach speeds upward of 54 miles per hour on long windward flats, but here under the low-roofed shop it looks incapable of rolling downhill with a push. We move closer and circle it, taking inventory. A pair of trouble lights hung from the stringers cast shadows that highlight every ding and dimple. Parked in my driveway under the light of day, the truck looked just generally shot, but here up close beneath the artificial light, the damage is more specific. Less theoretical. I had gotten in the habit of blithely telling people we'd have to patch up the one big hole in each fender. Now, poking my finger through the hole and running it around the sharp edge—a dangerous sensation like trying to fish a Cheeto from a pop can—I realize I have no idea how we're going to actually do the deed, or where to begin. I rest my hand on the curve of the hood and feel the spiky grit of the rust against my palm, at the same time imagining the satin feel of that curve after sanding and painting. Mark is tapping around the edges of the rust hole to see how far the damage extends. While I go golly-eyed for sensation, he is checking fundamentals. I can't see past the surface of the thing.

We move around to look at the grille. You could pull a dead cat through the holes rusted beneath each headlight. "Probably have to rebuild that . . . maybe with diamond plate," says Mark. I grab the bumper, which is bent but sturdy. "We should rig some sort of big old deer catcher," I say. "A brush buster!" says Mark. He's grinning. I feel good, having come up with something that pleases him.

So we stand there awhile then, framing up the grille, seeing the steel grid in our mind's eye, like painters sighting over one thumb. I joke that I want to attach a pair of those little deer warning whistles. We are caught up in the idea of the truck up and running. We are getting way ahead of ourselves.

○ ○ ○ ○

I am going on a date because I went to the library. Should the American Library Association wish to release that statement in poster form, they have my permission.

The Fall Creek Public Library of Fall Creek, Wisconsin, is fighting to survive the Age of Not So Much Reading. A week ago the library held a daylong fund-raiser featuring three authors, a man who plays ethnic flutes, a local woman dressed as Mary Todd Lincoln, a folksinger, and lemon bars. I had been invited to participate.

I didn't want to go. The library director had put in her request very early on, but I was coming off a five-month stretch in which I had been on the road more than off. In the four days prior to the event, I did five radio interviews, gave a magazine interview, and covered three ambulance shifts. Not exactly factory work, but I was short on sleep and sick of my own yapping. I had the first thickheaded ticklings of a cold. In short, I was nursing a case of the antisocial whinies. I called the library director and said I didn't think I would be there. Then, at the last minute, thanks to a pathological combination of guilt and sense of duty recognizable to Midwestern Scandinavian Protestants everywhere, I loaded my car with books and drove on down. Among other things, I have a vested interest in the survival of libraries.

The folksinger was just finishing up and I was due to go on when I realized I had forgotten a box of books in my car. I dashed back out into the freezing wind and up the sidewalk. A woman and a toddler rounded the corner by Big Jim's Sports Bar. As we passed, I gave them the cold-weather nod. The woman returned it. They seemed to be in a hurry. Later, when the reading was over and I was signing books, the woman and child passed through the line. While I signed the woman's book, the toddler—a little girl—peered at me over the table edge. I noticed that she had her mother's pale blue eyes. The woman paid for her books with a check, and on the check it said her name was Anneliese. This caught my eye, because the only other Anneliese I know is my cousin, who co-incidentally grew up in Fall Creek. I said something to that effect, the woman smiled, and I turned to the next person in line.

On the way out of town, I stopped to visit my aunt Pam and uncle Larry. Aunt Pam made coffee and we sat at the kitchen table. I mentioned the Anneliese coincidence. Aunt Pam said she knew the family. She thought this Anneliese had gone to school with my cousin Sukey the marine pilot.

○ ○ ○ ○

A few days later I received an e-mail from Anneliese. She said she was a happy Spanish teacher and parent of a three-year-old daughter. She said she was a small-town farmgirl reassimilating to the Midwest after time spent traveling and living in Europe, Mexico, and Central America, but then she also said she wasn't assuming that my "author's voice" and reading persona gibed with the day-to-day me. I took this to mean she would not subscribe to shuck and jive. She wondered if I might like to meet for a cup of coffee.

I e-mailed back:

What you saw and heard at the reading is the real deal, although as you might suspect, not the whole deal. The everyday unedited version is not all poetry.
But if you're willing, coffee would be nice.

Speaking of the unedited version, I didn't e-mail *right* back. First I conducted a preemptive Googling. I found a picture of her doing yoga and a request for a babysitter fluent in Spanish. That was pretty much it. My name is more common. If she had Googled me, she would have discovered that I was: Robert W. Woodruff Professor of Law at Emory University and a leading authority on the relationship of morality to law; a self-appointed expert on cocker spaniels ("*Dealing with fleas? Ask me!*"); Nellie B. Smith Chair of Oncology for the University of Missouri at Columbia; scenic artist for *Fright Night Part 2;* author of *The Groom's Survival Manual;* or, a board-certified sexologist from California whose "hot" products include the Love Swing.

I received her e-mail just as I was leaving for a speaking engagement in Milwaukee, so I ran it out the printer and put it on the passenger seat of my Chevy. It takes five hours to drive to Milwaukee. Every couple of exits, I'd steer with my knees and give the e-mail another read. Studied the nuance of every word. I have always politely declined reading-related invitations, mostly out of being chicken. The only reason I was considering responding to this Anneliese was because I suspected my aunt Pam had drawn on her connections to put the word out. About the time I was reaching Milwaukee, it struck me that my longtime friend and mentor Bob might be a source of due diligence. He had taught English in Fall Creek for years. I called him from the parking garage, asked him if he ever taught a girl by the last name of Scherer. Two of them, he said. Were they crazy, I asked, meaning in the certifiable sense. Not at all, he said. Wonderful girls. Well-read, world travelers, just wonderful.

Anneliese has asked me out, I said.

Oh, you must, he said.

$$\circ \ \circ \ \circ \ \circ$$

Mark breaks the reverie. He is standing beside the right rear fender, which is attached to the box with a semicircle of retaining bolts. "I want to see what it's like under there," he says, patting the fender. "We're gonna hafta puller." *Gonna hafta puller* is one of my favorite shop phrases. It applies in any circumstance where any mechanical object—the fuel pump, a bad spark plug, or the entire dang engine—requires removal. You say it with a tone of can-do resignation, and it helps if you take a big breath first and then speak like you're hiking your pants or lifting something heavy. I'm telling you, it really puts hair on your chest. I'll stand there sometimes with the hood up, looking into the bowels of the machine, and I'll just suck it up and say, "Yee-up, looks like I'm *gonna hafta puller*." And then I'll reach in and draw that empty inkjet cartridge right on outta there.

The bolts that keep the fenders tight to the sides of the box froze up with rust years ago. Rather than fight them with a wrench, Mark rolls out his portable plasma cutter, a compact unit roughly the size of a beer crate. The plasma cutter sends an electric arc through a blast of super-heated air, creating a jet of ionized gas that blasts from the tip of the handheld torch at just over 9,320 miles per hour. It cuts steel like paper.

The first thing Mark did when he got it was trim a sheet of eighth-inch steel into a silhouette of a miniature bear and present it to my sister. "Really," he says, "it looked more like a feral pig."

As Mark disappears under the rear end of the truck, I grab a sidewinder grinder, insert a circular wire brush, snug the chuck, and start buffing rust from the passenger side front fender. The rust evaporates at the leading edge of the whirling bristles, shooting off the fender in puffs of brown dust. When I pull the grinder away, the exposed steel shines silver as the day it was rolled. I run my fingertips over the shine. The steel is warm from the friction of the brushes. My finger pads turn orange with powdered rust. I keep caressing the clean spot, marveling at how close the fresh iron lies beneath the abrasive rust. Mark is out of sight, but I can hear the hiss and blow of the plasma cutter and smell the scorch of vaporized steel. Occasionally I hear clunks. Mark is under there cutting and struggling and I'm up here rubbing that shiny spot like Steinbeck's dim Lenny stroking a dead puppy.

After searing away the last of the bolt heads, Mark reemerges from beneath the chassis. I put the grinder down, and standing shoulder to shoulder we grab the fender and tug it gently, rocking it until it breaks free. The exposed studs are still hot enough to smoke your skin. We are looking at iron that hasn't seen the light of day since those bolts were tightened over fifty years ago. The rust has had time to work, and it's worse than we thought. With the fender removed, the rear wheel looks naked on the exposed length of skinny axle. Kathleen sticks her head in the door. She pulls night shifts as a box maker at a factory that, among other things, manufactures baby caskets. She tells Mark she is off to work. Sidrock needs watching, so Mark leaves for the house. I hang out in the shop and putter. Later, when I leave, I can see Mark through the yellow square of the living room window. He is on the couch with his back to me, head inclined over the baby in his arms.

○ ○ ○ ○

Along about 1990, I forsook all the tweaking and just let my hair grow. Long and straight. No fuss. For most of a decade it fell to my waist. I trimmed it back once or twice, wearing it short briefly, then grew it back. When I first started wrapping my pony tail in an elastic band, it was broomstick thick. Eventually I began to notice that the pony tail was thinner, until it was more the diameter of a pencil. The hair shafts were becoming brittle. Hair that once laid flat on the top of my head was developing twists and curls. I developed the creepy and hypnotic habit of staring into space while running strands of hair between my fingertips, plucking the ones that felt kinky. I have read that compulsive hair-pulling in a trancelike state is called trichotillomania and movie stars do it, too.

It is a matter of no little irony that once I finally quit trying to tame or inflame my hair and simply grow it, I began to lose it. My hair is still long, but I am bald or balding. The classification is utterly dependent on lighting. My hair loss is neither absolute nor complete. It is *under way*. I do not so much comb my hair as harvest it. The progression has been unrelenting, albeit mercifully gradual. By the time the loss became noticeable, I was pretty much comfortable in my skin. Pity the man who molts straight out of high school. Or worse, *in* high school. I was a few years behind a guy whose hair began thinning his junior year and was pretty much gone at graduation. We called him the Bald Eagle. The cruelties of youth are unstinting. He benefits from a long-term upside in that when I see him on the street a quarter century later, it strikes me that he has retained his youthful good looks. I have a brother five years my junior, and he went bald in his early twenties. At thirty-eight, I'm still a few follicles from Category Cue Ball. Oblique lighting still allows me the illusion of coverage. But when the light is direct, I can see the future, and it is shiny. I look in the mirror and think, Bald Man Walking. With apologies to Bob Dylan, I ain't bald yet, but I'm a-gettin' there.

For a man on the brink of baldness, delusion is hope: *The bathroom light is overbright . . . the mirror is substandard . . . the breezes of June have always blown this cool.* The man, after all, is grieving, and the putative first stage of grief is denial. Then the man spends the day announcing waterfights at Jamboree Days without a cap and for the first time in his life, sunburns his scalp, and denial shrivels like thin bacon on a cheap skillet. I can tell you that it took me a while to place the sensation. The crown of my head took on a tingly tautness, the sort of thing you might welcome in other bits of your epidermis but which scalp-wise feels like a combination of hives and paranormal possession. Holes in the ozone notwithstanding, you cannot equivocate a sunburned scalp. You are losing your hair.

Sometimes, just because I can, I stand before the mirror and flip my remaining long hair up and over my scalp in a comb-over—the baldness cure that dare not speak its name. I consider cultivating the Ben Franklin look, but all the frizzy top cover just makes me look like a meth-addled biker. Last month, business took me to a radio station in Eau Claire, Wisconsin. I was sitting in the lobby wearing a camouflage hunting cap. The morning drive team was doing their show in a glassed-in studio visible from my chair. Spotting the hair hanging down my back, the female member of the duo described me on air and cackled, "A mullet! Mullets are so 1985!" We could hear her over the waiting room speakers. For the benefit of the other people in the lobby, I remarked that poking fun at mullets was "so 1995." Later, I related the incident to a friend. I had my hat off. She laughed. "You don't have a mullet," she said. "You have a *skull*et!"

○ ○ ○ ○

Every year at this time, sometime after the snow clears and before the first green shoots appear, our little fire department convenes to burn the dead grass along the railroad tracks. The chief waits for a windless weeknight and pages everybody in. We put on all our gear and pull out all the trucks, then run around with propane torches, lighting the long dry weeds. It's a firebug's delight. Generally we start up by TJ's Food-N-Fun and let the flames work south. TJ will step out from cooking Tubby Burgers and watch awhile, and conglomerations of kids pop up at the street corners, many of them straddling bikes. Once the flames get clear of Main Street, some of us head south to Highway Q and start another bank of fire coming back north. We arm ourselves with rakes and water packs to keep the fire corralled and moving in the right direction. It's kind of a trail drive with flames in place of cattle.

Ostensibly, we burn the railroad tracks to eliminate the possibility of a freight train throwing a spark and starting a fire inside the village boundaries. It's also a low-key rehearsal for a real wildfire. But above all, it's a great palaver. Darkness settles early, and then you get that hanging-out-around-the-campfire feel. Faces flickering with reflections of flame, voices in the smoke. We work a little and talk a lot. This year we cover motorcycles, mortgages, health insurance, and trailer park rent. Someone tells a story about a gigantic homemade bong. "You had to stand on the landing to smoke it," says the voice, disembodied in the dark beyond the fire line. "Three people sucked the air out and the fourth person got a hit." Someone else wonders if it would be okay to needle one of the firefighters currently out of earshot about the pretty much verified rumor that his wife had kicked him out for good. "I think you wait on that until he brings it up himself," says another firefighter. Social graces, blue-collar style.

The flames flare and recede depending on what sort of fuel they're chewing. Clusters of bone-dry canary grass whirl up in a minitornado of sparks and tendrils of ash. The sparks flurry and slide across the black sky, extinguishing one at a time. We watch for flare-ups outside the burn zone, and stomp or spritz them as needed. The air is cool, and when I move in close to the flames, the heat feels good on my face. Suddenly one of the other firefighters rushes up and begins patting me. Normally, you don't get a lot of that.

"What are you doing?" I ask.

"Man, yer *hair's* on fire!" he hollers.

When it isn't falling out, it's bursting into flames.

The very next day, I call Dan at Wig-Wam Hair Fashions. Tell him it's time. He was booked solid for the week. "I can get you in next Friday," he said.

That'll be fine, I said.

◇ ◇ ◇ ◇

The first of my seeds have sprouted. By personal tradition, at first sight of cracked earth, I pull Dylan Thomas off the shelf and reread "The force that through the green fuse drives the flower." I have been told the poem is not about a seedling but rather about Dylan's tallywhacker. I ignore

this and read it as a gardening poem. The first few stanzas, anyway. By
the time I read the conclusion, I have to admit the tallywhacker analysis
has some truck. Freud was a bit of a nutball, but there's no denying that
from Aristophanes to Mojo Nixon, the tallywhacker has a rich history as
prevalent muse.

I have rigged up an adjustable rack of cheap fluorescent lights in the
south-facing window of my dining room where the sprouting trays get
as much natural light as possible. The fluorescents are hooked up to a
set of light timers that feed the plants light long after the sun has gone
down. I used to keep the plants and lights in a closet off the living room.
I thought it was a good idea because the furnace chimney runs through
there and keeps it warm, but in the end the heat dried things out too
quickly. I had a writer friend visit once, he was a committed smoker of
herb, and when he saw the electrical cord snaking into my closet and
the light glowing from beneath the door, he grinned and pointed and
said with great expectation, "Yeah?" I opened the door to reveal trays of
parsley and a couple of beleaguered leeks. His face dropped like the kid
who got socks for Christmas.

One of the germination trays has a heating mat beneath it. This is the
first year I've tried it. There are sprouts around the edges, but none in
the center, which seems to dry out quickly. The water condenses on the
little plastic dome, and runs back down the sides, creating a centrally
located patch of drought. I have some leeks going, also three sad little
basil sprouts, a clutch of lettuce mix, and some spindly parsley. I had
a cilantro sprig, but it gave up the ghost last Wednesday. No apparent
reason, perhaps it was my singing. Or overwatering. I tend to check the
seed trays two or three times a day to see if they've made any progress
since breakfast. I usually give them a spritz. Surely none of this helps.
Putting me in charge of seeds is like dropping your kids off with a weird
uncle who feeds them Funyuns for breakfast, then sends them out back
for an unsupervised game of Jarts. When you combine my time on the
road with my general gardening insouciance, you must expect a hefty fail
rate. Still, no matter how it goes, I never tire of watching these plants
unfold—slower than the naked eye can see, and yet, in day-to-day terms,
with amazing speed. And I love to snap off a scrap of something green

to eat while the ground is still hard. The chlorophyll dose bears you forward to warmer days.

○ ○ ○ ○

Although by and large we have not had much commerce, I like Dan the barber for a lot of reasons, including the fact that he parks up the street instead of right out front, which leaves the prime parking spot open, making things easier on the ladies coming in for church perms. I guess this is just good business, but still. I am also partial to him because when he came to town twenty-five years ago after leaving the Air Force and bought a tiny hairdressing shop called the Wig-Wam, he decided to retain the name, and believe me, I get some mileage out of that one while skipping through certain privileged circles.

But above all, I like Dan because every summer in the Jamboree Days beer tent we commiserate about our International problem. Dan has a Scout he has been trying to get refurbed and back on the road for several years now. At the moment it is in pieces in another man's shop. When I walk in the shop today, he greets me with alarm. "Did you sell your truck?!?" Imagine your best pal from the bowling team spotting your ball down at the pawnshop. I tell him no, tell him what's up, tell him as a matter of fact I'm going over to work on it later tonight. He is visibly relieved, no doubt in part because now we can suffer the same frustrations.

Dan asks me a couple of times if I'm sure about cutting all the hair off, but I'm unmoved by the whole thing. It's time, cut away what doth remain. So while he clips we talk, and watch the cars and people come and go. Dan's studio is elevated about four feet off street level, so he has this great vantage point, not real bird's eye, but high enough so people don't notice him observing. Sometimes I think of all the colorings and sets and rinses he's done, and how I bet sometimes when someone gets married or divorced or has a baby, Dan has seen it coming months or even years ahead of the rest of us. Sometimes I think if Dan blabbed, half the men in this town would need new nicknames.

"What do you think?" Dan has swiveled me to the mirror. He took it right down to an eighth-inch, like I asked him. I have forgotten that I have such a large round skull. Definitely got some Charlie Brown melon going there. But I rub my palm over it and it feels good. Now if I can just break myself of flicking back hair that isn't there.

When I emerge from the Wig-Wam, the air is still warm and I have time and daylight enough for a bike ride. I head east of town, over the county line into a stretch of roller-coaster topography. It feels good to get out of the saddle on the climbs, rock the bike, do the push and pull, pretend I am ascending Alp d'Huez, and not some hillock in Barron County, Wisconsin. Halfway up the first big rise I pass a swampy hollow. A peeper frog choir is in full swing. Most of the frogs are singing baritone, creating a cumulative low-end chuckle similar to diesels idling behind a truck stop. Somewhere in there a lone castrato works a high note that cuts through like a bell, rising insistently above the mix and not fading until I crest the hill and drop down the other side, where the land opens up before me. Everything is stark and brown. A few intransigent strips of grainy snow cling to the north-facing slopes.

On the return trip, I have the wind with me, and I launch from the hill with a burst of pedaling, then assume a tuck and really let it roll. I am nudging 45 miles per hour and whizzing past the froggy choir when a flock of wild turkeys explodes from the hazelnut bushes, crossing left to right. Upward of a dozen gobbling brown missiles fill the air ahead of me, above me, and behind me—everywhere but in the spokes. I consider the ignominy of being found dead with feathers in my bike shorts and a turkey beak jammed in one ear, but I keep my head and the hammer down on the hard-earned theory that evasive action generally compounds your problems. As I come safely clear of the turkey strafe, and the last of the frog notes yield to diminuendo, I am reminded once again that certain germane aspects of the naturalist experience cannot be conveyed via the pretty pictures on your National Wildlife Federation screen saver. I plane out onto a flat stretch, exhilarated at the ineffable cusp of spring, when frogs sing and turkeys attack.

On the last leg of the bike ride, I encounter a farmer friend and stop to visit on the shoulder. He's pulling a portable welder behind his pickup. "Rock rake's busted," he says, and that's enough. "Say, you sell that old truck of yours? I drove by your place the other day and seen it's gone."

"Nah, didn't sell it," I say. "Workin' on fixin' it. It's pretty shot."

"Yeah, we was talkin' about that down to TJ's. I said I figured your best bet would be to jack up the radiator cap and drive a new truck in under it!"

○ ○ ○ ○

I shower and make the drive to the shop. Mark wants to pull the bed so he can get it sandblasted. While he disappears under the frame again with his plasma torch, I go about removing the spare tire rack. First I grab a lug wrench to remove the tire itself. In the circle formed by the hole at the center of the rim, I can see the same hand-painted letters I've seen for years:

4–27-
Hi
Ron

When I pull the tire, the entire date is revealed:

4–27–82
Hi
Ron

Just a few days short of twenty-one years, then. I would have been a junior in high school. The twenty-seventh was a Tuesday, so I probably slouched around school, attended track practice, and then spent the balance of the evening in my bedroom listening to the theme from *Chariots of Fire* whilst longing for the farmgirls of spring. Who ran toward me in slow motion.

With the tire removed, I can get to the six bolts that secure the frame itself to the box. One of the nuts is missing. I soak the rest with WD-40 and then make two trips back and forth between the truck and the toolbox before I choose the socket sized to fit the nuts. Mark glances at them from ten feet and says, "Prob'ly five-eighths." Yes.

After a little handle-banging to break the rust, the nuts ratchet off easily. It helps that the rack is mounted on carriage bolts. Carriage bolts have a domed cap that give a wrench no purchase, but the shaft collar is squared off above the threads, which holds it from spinning when countersunk in a square hole, as these are. Rather than have to twist two wrenches at once, I just keep a little thumb pressure on the domed head, and flick the ratchet handle back and forth until the nut spins free. Ever since I was a kid helping Dad bolt down the bed on our hay wagon, ratchet sockets have been my favorite tool. The socket fits the nut so snugly. The hidden mechanism that allows the socket to hold its position while the handle is rotated backward for a fresh turn is a little bit magical. I've always enjoyed the *zip-clunk, zip-clunk* sound of the ratchet working, a sound that is simultaneously fun and industrious, and it's fascinating how such a little adaptation (the ratchet) amplifies the efficiency of the tool. When all five nuts are spun free, I set the rack aside and get my first good look at Ron's rendition of the Playboy bunny. Techniquewise, it's pretty well executed, with a bow tie and all, although it looks a little more Peter Cottontail than Hardcore Hare. Ron has enclosed the caricature in quote marks. You see that a lot in homemade signage, quote marks used for emphasis, with the end result being unintentional irony. As in "QUALITY" CAR REPAIR. Could it be that Ron—a man whose swabs of primer may have been inartful, but saved the truck from rusting away completely—is in reality a postmodern ironist? I have very little to go on—only a vague recollection of standing in his yard on a cold day and giving him some cash.

Mark has detached the box. We get on either side and try to shift it. It's mighty heavy, but not impossible. Using pry bars, sawhorses, and the brute force of our backs, we inch it up and rearward, increment by increment, until it is clear of the cab and frame and resting on sawhorses. The old truck looks suddenly lightened. The revealed frame rails—upon which the entire rest of the truck is stacked or hung—are surprisingly narrow and spare. So much of our impressions of automobiles and machinery is formed by the shape of the skin, you forget how relatively fine-boned the linkages are that hold it all together and make it run. The big rounded cab looks tick-fat and out of proportion to the twin rails of exposed channel iron supported by two narrow tires. The truck looks like a rooster with no tail. A fat man with skinny legs. But you get an idea of what the truck was designed for now that you can see the full length of the leaf springs—held to the frame with a stout shackle and bracket system and U-bolted to the axle in a thick stack, including a spacer upon which rests a stubby set of auxiliary overload springs for extra big loads. The setup looks so stiff you can't imagine the cargo that would make them flex.

Mark heads in to watch Sidrock for the evening and I putter a bit longer, bagging the nuts and bolts from the spare tire frame, making notes (*spare tire rack—one bolt missing*), picking up tools, cleaning up some. A few days after I answered her e-mail, Anneliese and I spoke on the phone. Tomorrow we will meet at a coffee shop. I tell her, Don't look for that guy with the long hair. I was wearing my scurfball camo cap when I did the reading, so I'm wondering if she knows how much chrome I'm sporting up top. But I'm not real nervous. I'm ready.

There was this pop song—"Drops of Jupiter" by Train—that peaked right about the time I took up with my last girlfriend. It is a beautifully overproduced musical tid-bit. In my teens, I would have wallowed in it. In my late twenties I would have sneered at it. In my mid–thirties— having been told by a wise friend that "there are no guilty pleasures, just *pleasures*"—I simply enjoyed it for what it was. The first time I heard it, I grinned and turned it up. There were strings, and longing, and a sweeping chorus, and just as I thought, *the only thing missing here is some na-na's*, the "*na-na's*" kicked in. I sang out.

The prevalence of this song coincided with the sweet harmony stage of the relationship, that initial stretch where you marvel at the align-ment of the planets while ignoring the fact that you are astride a falling star. When things went unignorably south, I really couldn't bear to hear "Drops of Jupiter." If it came on the radio while I was driving, I punched the scan button, hoping to snag a George Jones song. At home I listened to Tom Waits. *Closing Time*, mainly.

Then one day I was running errands, and "Drops of Jupiter" came on the radio, and I liked it again. The *na-na's* came and went, and my liver did not twinge. Put me back in, coach.

And so now I am in the car driving home in the dark nursing a quiet little blend of excitement and hope. God bless our unkillable hearts.

○ ○ ○ ○

I got to the coffee shop a little early, shaven and dressed in my favorite T-shirt, a black one that says ROAD KING across the chest. Karmic groove-wise, that is one of my top ten all-time T-shirts. The logo is encircled by stylized lug nuts. Wearing it feels like vitamins and valium. I wore steel-toed boots—the shinier of my two pairs. Jeans. Anneliese arrived minutes later and parked across the lot. She was driving a worn Honda Civic, black, with dents, a bike rack, and state park stickers on the windshield. She was small and blond and walked with grace and strength, and shook my hand with a smile that threatened to derail my objectivity. I held the door like a gentleman and, I cannot tell a lie, checked her out as she passed through. She looked delightful in her jeans. Civility is sublime, but humankind owes its existence to the animal urge.

We ordered tea and talked, nonstop and variously and with ease, as you do on any first date that comes up short of a train wreck, and then we had more tea and more talk, and then, it being a sunny day, we decided to go for a walk. By this time my bladder was distended to the point that my abdominals were creaking, and as I rose to my feet, I adopted a slight crouch in order to remain continent. Later she would admit to similar difficulties. We took a bathroom break.

It was a fine sunny day, and we walked and walked. I clomped along in the boots, learning more as we went along. She spent a fair chunk of her childhood on a farm, growing onions and sweet corn to be sold in town from the back of a pickup. She had gone to school with some of my cousins but she shot down my theory of Aunt Pam as matchmaker. There had been no contact. After teaching high school Spanish she was now teaching at the local university. We walked two miles. Then we stopped for Thai noodles in a strip mall, one of those places where you get rice in a pile the size of an orthopedic pillow and the sauce runs heavy to salt and fat. We ate and talked and then drank green tea and talked, until the sun dropped below the upper sill of the plate glass storefront and blasted me in the eyes, and then I walked Anneliese to the Honda, where she paused to stand with one hand on the roof and the other on the top edge of the open door long enough to say, yes, she'd like to see me again.

○ ○ ○ ○

Anneliese and I walked in the sun on Sunday, and by Monday the temperature set a record at 90 degrees. But it was a trick, a meteorological head-fake. The warmth was pushed in on the bumper of a cold front, which on Wednesday overtook us, precipitating a good inch and a half of rain. As the last few tenths fell, temperatures dropped to freezing and everything got a fat round coat of ice. This morning sunbeams flared from the trees in a million pieces and the clusters of precocious grass illumed the granulated snow in pale refractions of green.

Many of the seeds I started—including the oregano and sweet marjoram—haven't sprouted, so I head back in the basement at my gardening bench, sprinkling more seeds into more planters. I'm also repotting some leeks, cilantro, and two basil plants. Three basil plants had come up, but this morning I discovered that one had damped off and died. I am studying it now, pale and flat on the vermiculite, and I am thinking, *Why didn't you call? Why not wilt a little first, give me some warning? More water? Less? Should I have pulled your plastic cover? Lifted it less often?* Oh woe, and whither the pesto.

I drop the living basil plants on the garden bench four or five times in order to break the soil cube from the roots the way I saw Brian Minter do in a gardening video, then tuck them into new, larger receptacles. Then the leeks, and then the cilantro. I suspect the reason so many of the seeds I sow directly in the earth will catch and pass most of these presprouted plants lies in the fact that they are not forced to endure my fake lights and fiddling.

When I finish with the plants, I put them under the lights, clean up, and leave for my second date with Anneliese. I take my copy of *Waiting for Guffman,* and we watch it in the basement while her daughter sleeps upstairs. Anneliese laughs in all the right places, which I snootily think bodes well. My taste in films is largely nonexistent. When she says *Tommy Boy* is one of her favorite movies, I am so delighted I want to give her a noogie, but it's early. Then she says she and her sisters and

mother can recite every line in *What's Up, Doc?* and I admit I have
never seen it. We talk and laugh and watch all the DVD extras and talk
and laugh some more, and then, in the half-light of the foyer, I look
at her smiling up at me and thank her for a wonderful evening and as
I turn to leave a voice inside my head says, *If you don't kiss her right
now, you are a clod,* and I turn again, backtrack three steps, put my
hand in the small of her back, draw her to me, lean down, and kiss her.
She makes a soft little *mmmmm* sound that will echo in my heart until
my brain fades to black.

Half delighted and half panicked, I turn to leave and trip over the
doorsill, not falling, but stumbling onto the porch like a drunk. Composing myself, I walk straight down the sidewalk, where I misjudge the curb
and smack my knee on the car bumper.

Feels good.

$$\circ \quad \circ \quad \circ \quad \circ$$

If you're going to have a shop, you've got to have a shop radio and a shop
chair. Mark's shop radio is a beat-up boom box, which is nowadays an acceptable substitute, as long as the radio works. The thing about working
with the radio in the shop is that it should be a background thing, not a
dominating thing. The songs, the ads, the deejay, if they still have one,
they all bear you along through the hours. It's not about loudness or high
fidelity. A lot of farmers used to play the radio during milking, and part
of the comfort was the way the sound changed as you moved around
the barn. Mark is not a purist in this respect. He keeps a stack of CDs
in the shop: Metallica, Slayer, Pantera. "Nothing like a little 'Cowboys
from Hell' to ease your mind," he says with a grin. When I'm there he
usually compromises with Classic Rock, a favorite of neither of ours, but
acceptable for wrench work. Right now it's "Proud Mary" coming from
the shelf up in the corner there, the Creedence Clearwater Revival version, which is a good thing, because the Tina Turner version tends to
preoccupy me.

Meanwhile, the significance of the chair may be lost on anyone who
hasn't spent any time doing manual labor while standing on a concrete
floor. The simple relief of taking weight off your feet, of relieving the pres-

sure on your backbone, cannot be overstated. The chair should be comfortable, and it must be durable. One of the best I've ever known was a chair in a shop on a ranch in Wyoming. It didn't look comfortable—it was an all-metal, uncushioned minimalist beast—but it was constructed of spring steel, and when you sat on it, it sank down and back to the perfect angle of recline, cradling you with a soft bounce. My brother Jed has a vintage paint-spattered kitchen chair in his shop. The wooden legs are sawed off short, which gives it a nice low-slung feel, but the seat is split so if you're not careful you'll literally catch your tail in a crack.

Mark's shop chair is a heavy four-legged office model. I peg it to be a product of the late 1960s based on the fact that the steel frame is upholstered in burnt orange Naugahyde. The chair is padded enough to be comfy and sturdy enough to withstand rough treatment. I am sitting in it now, eating a fast-food burrito with my feet up before the fire. I have come to work on the truck late at night, as I often do. I like to drive through the country when the houses are dark, and work into the early hours when it feels as if the rest of the world is beneath the blankets. I usually call ahead during the day to let Mark know of my plans so he doesn't come creeping around the corner with a shotgun. Before he went in to bed, he started the fire for me. I am struck by the thoughtfulness.

As I eat, I peruse the J. C. Whitney catalog. If you're not familiar I'm not sure what to tell you other than if you place a Victoria's Secret catalog and a J. C. Whitney catalog before your average man, he will of course scamper off with the first, but later you will find him snoring on the couch hugging the latter. It defies brief description but "lots of cool doo-dads" will do. When we were boys we loved it, and now that we are men, we love it more. You can burn a lot of cash and quality time with a J. C. Whitney catalog, as I'm sure any number of marriage therapists will verify. It is hard to tear myself away, but the burrito is done. Crumpling the wrapper and chucking it in the fire, I harness the filter mask to my face, don my goggles, insert my earplugs, and set to scrubbing the truck. The sidewinder grinder proved to be overaggressive, so I've switched to an electric drill and sanding pads. I'm working on the passenger side door, and the pad generates a steady plume of erythmatic dust—the color a combination of the orange-brown rust, the pink primer, and the remaining genuine Harvester Red paint beneath. Even with the mask in place, I get this sweet ferrous taste back by my uvula that tells me some of the paint and rust particles are seeping through. The drill bogs a little, and the spinning wheel takes on a brighter tinge. When I lower the drill and inspect the area I've been scrubbing, the steel is smeared with gummy yellow paint. Some faint yellow has always shown through the primer on each door. I never gave it much thought, but really, this is a bit of a mystery. In 1949, the L-Line trucks came in four standard colors: Harvester Red, Adirondack Green, Apache Yellow, and Arizona Blue. Prior to 1949, the same four colors were known as Red No. 50, Dark Green No. 10, Yellow No. 165, and blue was not available. This disturbing trend would continue with the availability of optional colors including Black Canyon Black and Palomino Cream, and would culminate in the industry-wide disaster of those horrific factory graphics packages slapped on so many pickup trucks of the mid-to-late 1990s, which, combined with all the flare fenders, will one day render them the disco shirts of our age.

I am happy that my truck was originally Harvester Red, which seems honest enough and was in fact used to repaint International tractors when there was trouble with Tractor Red in 1953. But this yellow paint—for a moment I thought maybe it was the Apache Yellow and these were not original doors, but it's only on the main panel and doesn't extend all the way up to the windows or back to the trailing edge of the door, and the Harvester Red is still under there, so that's not it—must have been slapped on there by somebody sometime for some reason. I go around to the other door and find the same thing. I come back around and try a while longer to sand the first door, but the yellow paint just keeps gumming everything up, so I leave it and move to a fender.

I run the grinder late into the night. It is a good thing, to work with your muscles, to feel the grit on your skin and the vibration in your bones. The goggles and the mask and the earplugs create a silent core at the center of the noise and dust, and I relive the kiss from the night before, again and again, around and around.

○ ○ ○ ○

I wish I had thick beautiful hair the same way I wish I was six-foot-two with abs. I'm not, and so it goes. I'm okay with it, but you won't hear me buying into that whole bald-is-beautiful thing, which is the follicular equivalent of Size Doesn't Matter. You can only invoke Sean Connery so many times. And there are other problems. In the past week, I gouged my head on a protruding roofing nail in the garage and slammed my skull into a copper pipe in the basement. I did the usual head-clutching curse dance. I have been doing this sort of thing for years and no one the wiser. But this week two people asked me how I hurt myself. I was baffled until I got to a mirror. Without hair cover, two prominent scabs announce to all the world that I am terminally clumsy.

If a simple, safe, and affordable baldness cure comes along, I'm interested, and willing to hasten that cure by offering my remaining hair to science. In particular, three specific hairs composing a paltry little thicket just one-half inch south of my hairline. (Three-quarters of an inch, perhaps, by the time you read this, as in my case the term *hairline* represents the optimistic characterization of what has become a frankly hazy demarcation *in transitus*.) Like ship masts dwindling beyond the curve of the earth, the bulk of my hair is receding along the horizon of my bean, and yet this rebellious little trio stands fast, each strand apparently bald-proof. Why do they survive where so many have fallen? It strikes me that the holy grail of hair loss—a cure for baldness—may be woven in their DNA. I plan to alert the proper scientific authorities. Of course I hope to turn a buck. As such, I shall convene my accountant, my barber, my bioethicist, and a coven of intellectual property lawyers to compose a contract making all three strands available for lease, individually or as a package deal useful in conducting control group studies. Scientists in Pennsylvania recently announced that they were able to grow hair on a bald mouse using implanted stem cells. More than any time in our history, it seems reasonable to believe that a legitimate cure for hair loss is on the horizon, and my three hairs are prepared to provide the missing link.

As far as my cowlick, I recall it fondly. As of this writing, a little ves-

tige remains. For years I cursed the genes that set me up with a patch
of hair that ran against the grain, little knowing that complete follicular
failure was percolating in those same genes, and had been since the
moment I was conceived. From the get-go, my hair was programmed
to fall out. One is grateful this so rarely happens with the pancreas or
the eyeballs.

By the time I cut it, I wasn't fussing with my hair much anymore,
but now my beauty regimen is really streamlined. No need to use conditioner. No need to blow it dry. No need for combs. No hair in my eyes
no matter which way the wind blows. Six minibottles of motel shampoo
and I am ready to go for the year. I was raised in a fundamentalist Chris-
tian sect that not only frowned on vanity but viewed long hair on a man
as sinful. My hair experimentation was just that—experimentation, not
rebellion—but members of the church still saw it as a form of defiance.
I left the church long ago. In that sense, my going bald may have put me
on the path to redemption.

◇ ◇ ◇ ◇

I work the sanding wheel late into the night, clearing a larger and larger
patch in the fender, which gleams ever more silver. Now and then I
take a break to stoke the fire and rest in the orange chair. I wonder if I
am about to introduce another woman to my family only to have them
inquire after her fruitlessly a few months from now. I have never sur-
rendered to cynicism, and I don't intend to start now, but over the past
twenty-two years, certain patterns have been established.

I mailed a note to Anneliese today. Told her how much I enjoyed
the evening. Made a wry comment about the pad thai. I've always been
one for writing follow-up notes, probably because I'm so mumbly in the
moment. For all my failed relationships, I can say I have never written a
love letter I didn't mean.

And I have never written one I could bear to go back and read.

CHAPTER 6

MAY

I AM EAGER TO PLANT the garden, but my mother, wise after decades of fickle springs, tells me to wait. Let the ground warm a bit more, she says. For now my seedlings will remain under their lamps.

It is still early days with Anneliese, but so far, so good. I went to her house for a sit-down dinner with several of her friends. While the food was being prepared, I joined Anneliese's daughter Amy in the living room. We set up a corral on the carpet and she instructed me in the choreography of plastic horses. I threw myself into the role, making clip-clop noises and nickering in the way I thought a plastic horse might do. Things were going swimmingly and we were smiling broadly when Amy froze her horse in mid-canter and said, "You have funny teeth."

Well, fair enough. I have a gap between my incisors that runs about an eighth of an inch. Officially, this is called a diastema. Madonna has one. So does David Letterman. Ditto the astounding Lauren Hutton. Sometimes I wear a T-shirt that says, I'M GAP-TOOTHED, AND I VOTE! and you better believe it. Because of the diastema, I can spit water like a human Super Soaker. When the time is right, I will show this trick to Amy and perhaps win her love forever.

As far as the community is concerned, Anneliese and I had our official coming-out party as a couple when we dined together at the recent New Auburn Area Fire Department annual chicken barbecue dinner. Ostensibly a fund-raiser, the chicken dinner generates a few hundred bucks for our department every year, but the main purpose of having it is to have it. Along about mid-morning we drive all the rigs out of the fire hall and park them in a row along the street so folks can gawk at them on the way in. Next we take to the empty truck bays with a garden hose and push brooms, sweeping up the grit and sluicing it down the cast-iron drain grates. When the floor is clean we rummage around in the storage racks at the back of the hall until we find the spray-painted plywood signs that say CHICKEN B-B-Q 4–8 PM FIRE HALL. A couple of the guys toss the signs in a pickup and run around town planting them at strategic intersections—one out by the freeway exit, one by the stop sign just past the railroad tracks, and one up where Elm Street hits Old Highwy 53, just up from the fire hall. We used to put a sign in the parking lot of St. Jude's to catch all the Saturday-afternoon Catholics, but since Bishop Burke closed the church, that demographic is lost. There was some pro-test over the closing and even television cameras, but Bishop Burke is not one to trifle. I went to the final service just to show solidarity for my Catholic friends—the ones who tithed this modest building into exis-tence and regularly opened their doors to people of other faiths when room was needed (including my brother, on the day he buried his first wife). Bishop Burke looked a little pinched, but he did not waver. That boy is going places.

Next come out the folding tables and chairs. Over in a building on Pine Street they have a machine that converts plastic beads into plastic tablecloths for Wal-Mart, and they give us as much as we need, as well as cardboard barrels for the trash. We tape the tablecloths to the underside of the table, and set each table with a bowl of sugar packets and creamer, as well as a salt-and-pepper set, the ones sold in a shrink-wrapped two-pack and fitted with plastic tops that allow you to select "pour" or "sprin-kle" with a twist of your fingertips.

Every year, the rookie lowest on the roster has to scrub the toilet. This year it's Ronnie. We wait until he's in to his elbows, then one of the guys sneaks in behind him and snaps a Polaroid with the camera we use to document accident scenes. As soon as the picture develops, we tape it to the white board in the meeting room, where it will remain on display until someone new joins the department and bumps Ronnie up a rung. Later, when Ronnie's wife arrives, we will make sure she sees the picture, so she knows he is capable of such a thing.

○ ○ ○ ○

After the dining area is set up, we start cleaning chicken. For the sake of sanitation, we use the medical gloves from the rescue truck, so everybody has blue hands. The chickens arrive plucked and quartered, so "cleaning" basically consists of giving them the once-over and thumbing out the kidneys. This year we went through four hundred and fifty quarters and a bunch of drummies for the kids. The meat is only partially thawed, and halfway through the job my thumb tips ache with cold. I am working beside Bob the One-Eyed Beagle, and because he is a butcher, the cold meat doesn't bother him at all. We work standing around a rectangular folding table and joke and yap and flick each other with kidney bits.

A couple of the guys on the department have rigged up a barbecue trailer. It's basically a steel box the size of a small Jacuzzi set on wheels and fitted with a hitch. Tim, one of the designers, backs it around behind the hall, parks it just outside the back door, and levels it with the trailer jack. Every year Tim and the chief argue about how much charcoal is required. And every year it winds up being eight bags, one dumped in each corner, two in the center, and two bags added later. After the application of enough lighter fluid to defoliate Rhode Island, whoever has the charcoal lighter sticking out of his back pocket sets each pile aflame.

While the charcoal burns, we pack the chicken on three grates made from industrial-strength expanded metal. The grates are hinged so they shut like a book. You put the chicken on one side and then fold the other side over and latch the handles. The chicken is squeezed tight so that when the grates are flipped back and forth on their center pivot, the quarters stay put. If we arrange them carefully, we can get forty-five quarters on each grate.

When the briquets are flickering white, Tim uses a shovel to spread them evenly across the bottom of the trailer. Then we lower the grates in place. The chief grabs an industrial-sized lemon pepper shaker and douses all the quarters, and then we settle in for what is always the best part of the day—the tending of the chicken.

For the most part, this is a hen party for roosters.

We have a pretty good chunk of cooking time before the crowd arrives. Some of the guys smoke, some dip a new chew, some nurse a soda, some stuff their hands in their pockets and just stand around. We fiddle with our ball-cap bills and tell stories. There was the time I roared up first on scene with the new brush rig at a cornfield fire and then couldn't get the pump to start. Finally an untrained bystander reached in and turned off the kill switch. Eric remembers how he and I were making an interior attack on a barn fire when I spun around and hollered, "I think I left my teakettle on!" When he wound up running a tanker shuttle into town, he swung by my house and let himself in the back door. Sure enough, upstairs in the bedroom where I do my writing, the hot plate was on, the kettle was starting to scorch, and the plastic whistle had melted.

The story stick passes around. Ryan fell through the ice on his way to go fishing last winter. He went all the way to the bottom and bounced back to the top without losing his fire pager. Now we call him the human fish finder. Ryan in turn reenacts the night Bob the One-Eyed Beagle bent over to pull up his bunker pants and we all saw the hearts on his boxer shorts. Every now and then flames spring up from the briquets and the chief knocks them down with his trusty squirt bottle. It's what a fire chief does. At regular intervals, we flip the chicken.

We keep a big steel pot of water boiling off to the side on an LP burner. The pot is filled with gizzards and sliced onions. Every now and then you jab a gizzard, dash it with salt and pepper, and then eat it. It's hard to wait for it to cool, and you wind up chewing in fast breathy bites.

This year the chief is armed with a digital meat thermometer and he keeps jabbing the chicken. "It ain't done," says the chief.

"It's done," says Tim.

"Hell if it is!"

"Hell if it ain't!" And so on.

"Guaran-damn-teeya, first person brings their chicken back 'cause it's red on the bone, I know whose ass I'm gonna chew."

We're already unloading the rack, using steel tongs to snatch the quarters and drop them into insulated coolers lined with butcher paper. The meat is seared with the diamond pattern of the metal grates and you can't hardly stand not to pick at it. Tim pulls a quarter apart to test it. We all stab our hands in, stripping off pieces. The plain meat tastes so good in the open air. Even better are the bits of chicken skin stuck to the grate. Crispy, greasy, smoked, and tangy with lemon pepper. At this point we'd be happy to take down the signs and eat the whole works ourselves. We pick the steel clean. Then we put on fresh gloves and pack the grates with chicken again. There is the age-old question of why men who won't heat their own macaroni water will nonetheless hump forth to grunt and wave sharp sticks over meat cooked al fresco. Ours is an emancipated department, recently hovering near a dead-even fifty-fifty distribution of men to women. And yet, elements of the chicken feed tend to break down along the lines of sex. Each member of the department is expected to provide a tray of bars for the dessert table, which is set up against the wall in the space usually occupied by the brush truck. The table is stacked edge to edge with desserts. Not a single one baked by a married man. This is nothing new. When Jimmy Smith joined the department a year after his wife, Brianna, he said it was because he was sick of making brownies. Of all the bachelors on the department, I am the only one who baked his own tray of bars. The other single guys just brought store-bought cookies. Which I might as well have done, as my bars are inedible. The recipe was simple. Oatmeal bars with chocolate frosting. Mix oatmeal with Karo syrup, spread on a pan, bake. Not sure how you screw that up, but I did. The syrup turned so hard it is as if the oatmeal is set in amber. I frosted them anyway. It was like spreading spackle on chipboard. I had to use a cleaver to cut the individual squares, leaning all my weight on the blade and rocking back and forth. Then I plated them under plastic wrap and made a show of putting them on the table. The important thing here isn't that you *make* bars or that people *like* your bars, it's that your fellow firefighters and auxiliary members see you coming through the door *with* bars.

I mentioned the need for bars to Anneliese the week prior. She detected my hopeful tone, and did not go for it. "Our relationship," she said, "has not reached the stage where I make your bars."

○ ○ ○ ○

While the second batch of chicken is cooking, I sneak home to straighten up the house in preparation for Amy and Anneliese, even running a vacuum over the living room rug. When I get back to the hall the rescue truck is just returning from a medical call. I didn't hear the pager over the noise of the vacuum. I make the mistake of saying so. "Why are you *vacuuming*, Mikey!?" asks the Beagle.

Big laughs all around.

As soon as the second batch of chicken is done, I sneak home one more time to shower and put on a clean department T-shirt. When I get back this time, the brush truck is just returning. This time I missed a small grass fire. The pager went off while I had my head under the shower. Remarking my second missed call in a row, the Beagle says, "Mikey, yer gettin' *all* tuned up!" More laughs.

We set up the serving benches. Napkins, plastic silverware, and Styrofoam trays to the right. Then buns and butter. The chief's wife has made gigantic tubs of cole slaw and potato salad, which we'll transfer to serving bowls and ladle out with ice-cream scoops. We have two Crock-Pots brimming with pork and beans. The chicken in the insulated coolers cools if you flap the lid, so we serve it from Nesco roasters. The Nesco roaster is as essential to community functions as folding chairs and willing citizens. It is the centerpiece of family reunions and church suppers and graduation parties. I like the older version best, all overbuilt and sturdy, a cross between a chamber pot and a bathtub, an enameled white Buick of an appliance. Lately here they've sort of leaned out the design, adding contemporary graphics and a ventilated handle, and I intend to write them a letter. Much of the comfort in the classic Nesco roaster derives from the old-school sturdy look. It is the tugboat of comfort food, and should not be fiddled with.

We place the drinks table just inside the doorway leading from the serving bench to the big truck bays where the dining tables are. Another folding table, two more coolers, one filled with water, one with punch mixed by the gallon, and two big thirty-cup aluminum coffeemakers flanked by stacks of Styrofoam cups bearing the logo of a local insurance company. For the cold drinks, we have plastic Budweiser cups left over from the Jamboree Days beer tent. Between the roasters, Crock-Pots, and coffeemakers, we always blow the breakers. While the chief cusses and punches the reset, four of the guys rustle up extension cords and braid them every which way, spreading the load so we can eke by.

I am still helping with the setup when Anneliese arrives. Amy is trailing behind her, weepy because the training wheel on her bicycle has loosened. Tim, who has three girls of his own, grabs a wrench and fixes the problem. He is a big man, and Amy looks at him with reservations until he puts her back on the seat and she pedals away smiling. I can see Anneliese is talking to a couple of the firefighter's wives. It's cool and overcast, and she must have stopped by the house, because she is wearing my old flannel shirt. There is no better way to quicken a man's breath.

The first trickle of customers arrive, mostly elderly couples. Lieutenant Pam takes their money at the folding card table and makes change from the tin box, then they move around to the serving bench where each person picks up a Styrofoam tray. Most of the men wrap their plastic fork, spoon, and knife in a napkin and tuck the package in a pocket—front of the dress shirt for the older generation, back of the jeans for the younger. A few "lakers" up from the city call their order in, then pick it up to take back to the cabin. A pair of sisters who live in the retirement village call on the phone, and we send someone over there with the order. Shortly the trickle picks up speed, until before you know it the line is wrapped around the garage and the walls echo with chatter and the scrape of chairs.

We serve wearing the blue gloves and white aprons and our clean ball caps with the fire department logo at the front, and at one point Ric and Ryan put the gloves over their heads, pull them all the way down over their noses, and blow them up. The five fingers stick up in the air, and the boys look like a cross between the Blue Man Group and cyanotic chickens. There is unabashed clowning, with everyone guffawing. But beyond that, it's such a civilized moment, handing the food across. A simple transaction, a certain respectful order, and all in all, good food. From behind the bean pot, you see the faces slide by left to right, and you feel a mix of recognition and rediscovery. The year's changes, passing by like a filmstrip. I especially cherish when some of the long-timers like Mrs. Jabowski come through, because they evoke the little boy in me. I feel deferential and straighten up a little, and go home thinking I should live a little better if only because Mrs. Jabowski is watching.

We take our own turns eating. I eat with Anneliese and some friends from up the road in Chetek. More simple, good moments. Simple food, simply prepared, paper plates and plastic forks with your friends and neighbors, no pretense. You are struck by the privilege of it—the idea of gathering peaceably, without fear, to feast. Our Annual Jamboree Days celebration always feels like a combination class reunion and family reunion with beer, and I haven't missed one in ten years, but the chicken feed is my favorite in a quieter way. Here, we are just getting together for dinner.

I kept an eye on my frosted oatmeal bars. Two or three disappeared, but then word got out, and the rest remain untouched. When Anneliese and I walked home that evening, I carried the plate. Instead of cutting through by the Legion Hall we took the long way around Tugg's Bar so Amy could ride her bike, and it was sweet to stroll hand in hand with my girl in my hometown, but later when I offered her the bars to take home she said no thanks.

○ ○ ○ ○

I have a young friend named Adam visiting this weekend. Yesterday he went fishing with my brother John, and today he has come with me to work on the truck. Adam's mother and I were in a relationship for some years, and then there were changes and a move, and now his mother and I are still in a relationship, only now it is a cordial friendship over some distance, and Adam and I see each other only a couple of times each year.

When Adam and I arrive at the shop, we find Mark already at work, knocking the rust off the frame rails with a pneumatic needle gun. Powered by a high-pressure air line, the needle gun consists of twenty-some eighth-inch-diameter rods arranged in a compact circle. A piston behind the rods rotates at high speed, essentially creating a swarm of miniature jackhammers, chipping and pummeling the rust from the underlying steel. It is a blunt-force tool, unsuitable for the relatively delicate stock of fenders and door panels, but perfect for the heavy channel iron of the frame. The tool sets up a deafening racket, so when we arrive Mark sets it aside and busies himself with removing the rear wheels. He's going to tear down each hub, pull the axles, check everything out, clean everything up, and put it all back together with fresh lubricants. Earlier he unscrewed the plug from the base of the differential case—popularly referred to as "the punkin"—and about four tablespoons of oil ran out. "Should be about a quart and a half in there," says Mark. He's been reading the specs in the original *L-Line Motor Trucks Service Manual* I bought off the Web. "Then again, it's not like you've been running at high speeds."

Adam and I turn our attention to the front of the truck. A friend of Mark's stopped by the other day and ground the entire rear surface of the cab—easily half a day's work—but large swaths of rust and primer remain on the hood, fenders, and cab crown. Adam and I are going to start by attacking the gummy yellow paint on the doors. First I swab them with paint stripper. The stripper goes on as a clear, thin paste, and almost immediately the paint beneath it softens, ripples, and begins to slough. The magic of nasty chemicals. We let the stripper work awhile, then I give Adam a putty knife and he begins to scrape away the residue. It is slow going, but for the most part, the gummy yellow paint comes away neatly, wadding up on the knife like toxic butter. The doors will still need to be sanded, but the going won't be so gooey.

While Adam and I were working on the door, Mark removed the right rear tire, but now I hear cussing coming from the left rear. He has made several trips to the toolbox, and now he's dragged out the impact wrench. I ask him what's up and he says he can't break the lug nuts loose. He's got them all dosed up with penetrating oil, and he's been reefing on the lug wrench with a cheater bar, but still, he says, they won't budge.

Like the needle gun, the impact wrench is powered by air and generates tremendous force. He is about to slip it over the first nut when I remember something I learned twenty years ago in Wyoming.

"You try turning them clockwise?"

Silence. He's looking at me like he may have to put me out of my misery.

"Some of your older model vehicles ran left-hand threads on the left-hand side."

He's still looking at me like he may have to chuck a socket at my noggin, but he kneels down and slips the lug wrench over the nut. Puts his weight on the handle and bounces a little. There is a sharp squeak, then his shoulder sinks as the nut eases loose and the wrench handle rotates to vertical. "I'll be damned," he says.

If he had hammered away against the threads with that impact wrench, he'd have likely twisted the studs off flush. "Then we'd'a had some issues," he says. He's grinning. We all grin. Me, I am unreasonably tickled that I have had a chance to teach the teacher. Score one for lefty-loosey boy.

Then we get back on task, and soon there is no talk, just the radio, and the dog coming by now and then, nosing hopefully at her plastic ball.

. We all work quietly, and I spend most of my time thinking about Adam, working out of sight there on the opposite side of the truck, scraping the curdling paint from the truck door. When I met him he was a little boy, and now he is on the verge of his teenage years and their concomitant turmoil. Soon weekend fishing trips and puttering will not hold his attention. He has always had a sweet heart, this boy, but it has not been easy for him, with his mother the one steady presence in his life. He is a veteran of custody wars that recognize no truce, and he knows what it is to be caught in the crossfire of all but bullets. We have had some rare sweet days, he and I. I have seen him strain to hold up a fish as long as he was tall, and him with a smile equal to the task. We have tromped through river bottoms and over Dakota Badlands, we have watched fireworks over the lake, we have made up silly songs about cows and bologna. I have put batteries in his Christmas toys, made him finish his salad, and there were nights after long drives home from Grandma's house when I carried him sleeping to his bed. Sometimes his smile and bright spirit were the only things buoying his mother and me through days of rage and spit. He has given me moments that will forever inflect the lexicon of my heart. And now all I can think is that I have failed him, that I am one more man who moved on, one more man who proved nothing except that his mother alone is true. The term *broken home* is inept and inapt, but I am the carpenter who came and went, making a few small repairs but leaving the big job unfinished. Tender intent has no more relevance than the wake of a sinking boat.

The moments pass, everybody silent for a while, everyone working with their hands and their own thoughts. We put in a good session, just three guys tinkering, sometimes mute, sometimes shooting the breeze. Adam gets the door scraped down. When we leave Mark to drive Adam back to his mother, the back of the stripped cab is glowing dully, and the wheelless rear end is propped up on a pair of floor jacks, a truck without wheels, getting ready to roll.

○ ○ ○ ○

On the second Wednesday night of every month we have our fire department first responders meeting. We review the previous month's calls, check equipment, and then have some sort of continuing education. After the meeting I head over to work on the truck and pull into Mark's driveway at 10:15 P.M. He has done a lot since my last visit. After turning on the yardlight, he leads me to his trailer. It's parked beside the shop and covered with a tarp, which is weighted down with rocks. We pick off the rocks and Mark rolls the tarp back to reveal the box from the International. It's been sandblasted. All the old paint and rust is gone. Unlike the fenders, which shine like new when I scrub away the rust, the sandblaster leaves the steel looking dusty and flat. If you back off you can see the pattern left by the nozzle. The good thing about sandblasting is it leaves the surface roughed and clean, the perfect surface for absorbing and holding paint and primer.

Next, he leads me inside the shop and, with a big grin, points at the new exhaust system. Sometime in the late 1980s, I pulled away from a stoplight at the intersection of Eddy Lane and Highway 53 in Eau Claire,

Wisconsin, and the muffler fell off. I liked the way the engine roared, but it was frankly deafening and, furthermore, when you drive a dilapidated bucket of dangerous junk to begin with, you don't need to attract closer attention by blasting the cap off your local traffic officer. So I went to Farm & Fleet and got a tractor muffler. Sadly, when the old muffler fell off, the mounting brackets went with it, so I had to secure the new muffler with the wire from two coat hangers. It worked okay, although sometimes when I hit really big bumps, the muffler would separate from the header and the truck would suddenly go loud. I'd have to pull off the road, put on a pair of leather gloves, and jam the muffler back in place. I had forgotten about this until we pulled the box last month and I saw the old coat hanger wires still twisted to the frame.

Mark has replaced the muffler with one that actually fits, and, using flexible exhaust tubing, he has extended the exhaust outlet so that it protrudes from behind the right rear of the cab, where he has fashioned a shiny stainless steel tailpipe. The tailpipe emerges at an angle and is cut on a bevel. It looks positively racy. We just stand there and grin at it for a while. The iron frame rails and crossmembers are dark and oily. After knocking off the loose rust and brushing them clean, Mark has coated them with rust converter, a concoction of tannin and organic polymer that converts voracious iron oxide to stable iron tannate. He has also mounted new shocks. They are shiny and red.

Kathleen is working again, so Mark has to get back in the house, where Sidrock is sleeping. He's been colicky, says Mark. "I never used to know what colic was," he says, as he heads out the door. "Colic," he says, over his shoulder, "is when both baby and momma are crying." I am smiling, because it's been interesting to watch the transition since the baby came. Playpen in the shop. Car seat in the big four-wheel-drive pickup. One night in the shop Mark told me, "I used to sleep and have money . . . now I have a baby." He's smiling, but he doesn't sugarcoat it. The early days were tough, he says. He knows he's supposed to speak in terms of joy and fulfillment, but the truth is, it's been a rough go. Still, the little boy already shows an affinity for wheels and tools, and the light in Mark's eyes when Sidrock skids around the shop in his walker and gargles at the truck chassis makes it clear he has found his joy. Happens to the best of us, I tell him. I once worked for a man in Wyoming who roared around in a rumbling four-wheel-drive Ford with chrome side pipes and monstrous fat tires. The year I hired on, he had just gotten married. The second year I showed up to work, the chrome pipes were gone and the tires had been replaced with the regular skinny version. The third year, I found a diaper on the dashboard.

I don the safety glasses and the hearing protectors and start the old familiar grinding session. I grind away until 2 A.M., getting most of the large flat surfaces of the cab done, including the roof. Up along the brow of the cab I find a series of pinholes. Strange, not sure why the rust would attack in such a way in such a place. Condensation, maybe. Tonight while I grind I am thinking about the past, and my gut is queasy. I'm wishing I had met Anneliese ten years ago. Before we had so much history. It's an infantile internal tussle between selfish and silly.

◇ ◇ ◇ ◇

I have spent the morning digging in the raised beds, turning the earth with a potato fork, shaking out the root webs and mixing in dirt from the compost pile. In the afternoon I make a run to my brother's farm and bring back a load of pigpen dirt, which I use to fill a large new raised bed. I have to haul the pigpen dirt in a trailer behind my Chevy. This is the kind of chore that makes me want to finish the truck, which I have lately only been getting to in fits and starts. I am always coming and going, always working deadlines, always doing things as they absolutely need to be done. No matter our vocation, we so often find ourselves living life as a form of triage. I need more time with the dirt, the scents of the soil with its inferences of dust and mud, of drought and plenty.

My mom gave me some tomato plants, which I figure means I can go ahead and plant them, so I put them to root in the bed with the pigpen dirt. Out in the backyard, in the raised bed that spends the most time in shade, I plant some leaf lettuce and a mesclun mix. It's a small start, but it's good to have something in the ground. It's nice to think that while I am running hither and yon, those tomato plants will stay in one place and grow.

I get to the shop early enough to work with Mark for a while. I have a blister on my thumb from shoveling the pig dirt. It's warm enough that we have the overhead doors open. Out across the Blue Hills the leaves are out and green, but not yet fully flush. Tonight we are listening to country music, which means Mark is being polite and surrendering the dial. He is not a fan of the ol' twang-and-weep. Before we get started, I have a little surprise for him. I've been using a couple of disposable cameras to document our progress on the truck, nothing fancy, in fact a lot of the photos are too dark to see much. One of the cameras had been knocking around in my car for a couple of months now, and when I finally took it in, the first several pictures were ones shot way back in January, when the truck was buried in snow. They were taken on a bright clear day, and for some reason—the angle of the sun, perhaps—the rust was highlighted in such a way that a distinct pattern was revealed on each door. I recall seeing strange lobes of rust on the door before, but this picture showed that I had been tricked by an optical illusion on the order of those little wooden signs that look like a line of sticks until suddenly your brain flips a switch and the sign very clearly says "Jesus." When I looked at that photo, the switch clicked, and suddenly the lobular shapes went to the background, and what leapt to the fore were big fat flames. The yellow paint! Sometime before Ron swabbed the primer on, he or some other wise guy had painted stylized yellow flames on each door. I show Mark the picture and he cracks up. We agree—it's like putting racing stripes on a hippo.

Tonight my job is to remove the steer wheels and pull the bearings. At the back of the truck, Mark is installing new brake lines. He is amazing to watch, weaving and bending the tubing. He has this ability to see things, to work with metal and mechanics. This reflects time and study and years of hands-on, but surely it is at some level innate. He is working with raw stock, and in order to make the fittings, he is using a double-flaring tool. It's a simple two-piece clamp-and-die device that, when tightened with threads, flares the end of the brake tubing stock to help it form a union. It's such a simple gadget, and yet it does a profound thing, reshaping a steel tube to give it function. Looking around at all the stuff crammed inside these concrete block walls I am reminded again that the dustiest shop is jammed with testaments to human ingenuity. At some point in the progress of homo sapiens, someone among us invented the double-flaring tool, and now your brakes work.

Because the front end of the truck is nosed up into the dark corner of the shop, I string trouble lights around so I can see the lug nuts on those CO-OP Country Squire tires, which look as if they came standard in 1951 and have never been changed out. The nuts break loose readily, and soon I have the tires off. Before we started this project, you could see the two obvious rust holes in each front fender. When I ground the right front fender the surrounding iron was fine, but on the left front the grinder revealed a myriad of pinholes similar to those on the cab brow. I remember my brother Jed saying it isn't unusual to see more rust damage on the driver's side of Midwestern vehicles, because that side is always getting splashed with slush and road salt from oncoming traffic. The trouble light is hanging directly overhead now and when I stick my head inside the left fender it is like entering a miniature planetarium. Little constellations of light coming through everywhere. This is not good. It's one thing to patch a big hole, but when a fender is laced this badly, you have to come up with some other solution.

I hear cussing from the back of the truck.

"What?" I ask. Mark has a measuring tape out and laid across the end of the freshly flared tube, which has been refusing to seat.

"Five and a quarter?!? It should be five and one-sixteenth!" He cuts the flare off and starts again. I make a mental note to spare Mark my Rhapsody on a Flare Tool.

Kathleen brings us a bag of cheese curds, and we dip into them. They're fresh, with the squeak. Your quality curd should squeak between your teeth. The squeakless curd is still tasty, but it is not prime. The curds pick up a little black grease from our fingers, but if you eat curds while you're wrenching, they're bound to get grubby. Grubby is comfy. Grubby says you're working close to the ground. That you don't have far to fall. Kathleen lingers awhile, visiting while Sidrock chortles. We tell her about the yellow flames, and I tease the dog by making it chase the flashlight beam. These are the moments I long for when I am running too far too fast. These are the moments when time slows and it is easy to breathe.

Kathleen takes Sidrock in to bed and Mark and I keep working as the light fades. Now that the front wheels are off, I start in on popping the grease cap that covers the end of the steering knuckle spindle. The cap is shaped like a small shot glass or an overlarge thimble, with the rim fitted tightly inside the bearing housing. I coax it loose by tapping outward with a plastic hammer. Finally it pops free, revealing a crenellated nut kept from spinning by a cotter key. I pull the key, spin the nut free, and then pry loose the tapered bearing, which is black with grease that smells dark, rich, and fatty. Out where the bearing rides, the naked spindle is less than an inch in diameter. I'm reminded again of how insignificant the frame looked when we pulled the bed. When you're bombing around out there, you forget sometimes about the relatively delicate construction of the pegs you ride on. I wander off to look for some solvent, to start washing things up.

When darkness falls, the June bugs come, motoring through the airspace of the shop, their buzz loose as a playing card in bicycle spokes. They are attracted by the trouble lights, in this case very aptly named. The June bugs are the size of double-wide kidney beans, and when one of them hits the halogen filament you hear a sizzle and then a plasticky clack as it hits the floor. Some miss the light and smack the concrete walls. They, too, fall to the floor, where they scrabble about on their husklike backs, unable to right themselves. June bugs have design flaws.

Having finished with the bearings, I start pulling the cast aluminum lettering from either side of the cab right above the fenders. In two parallel strips the lettering says INTERNATIONAL L-120 SERIES. (The spec plate inside the cab says my truck is actually an L-122; International offered the L-Series in sixty-six different models—depending on who you ask this reflected a commitment to customer choice, or a lack of focus that ultimately helped doom the company. Regardless, it can be confusing.) Each strip is attached with spring-loaded tabs pushed through holes drilled in the steel. In order to get to them I have to back into the cab and basically sit upside down on the seat, feet hanging out the door and torso bent backward so my head is looking up under the dash. By pushing against the tabs from inside with a piece of steel rod, I can create just enough space between the lettering and the quarter panel so that I can pry them loose with a flathead screwdriver. I ease each section off slowly, working end to end, not wanting to snap the delicate cast aluminum. I get them all out okay. While I'm in there, I clean out the glove compartment and find a to-do list (printed up the last time I tried to get the truck running), an expired registration, an ice scraper, and a copy of Louis L'Amour's *Catlow*. Mark has figured out the flaring tool, and now he's sitting on a crossmember between the frame rails with the manual on his knees trying to figure out where the rear brake line goes in relation to the punkin'.

I head home. I need sleep. Tomorrow I am to give the commencement address at my old high school. I've been kicking the speech around in my head, but at some point I should get it in print, double-spaced, like my English teacher taught me.

$$\circ \; \circ \; \circ \; \circ$$

I did the best I could for the New Auburn High School Class of 2003, and I remain honored by their invitation, but they are to be excused if they felt I simply delayed delivery of their diplomas by ten minutes. After prefatory remarks including the story of how I recently lit my hair on fire (it seemed relevant), I cut to the chase:

Class of 2003, I really don't know what to tell you. I graduated from New Auburn twenty years ago. With every passing year, I feel as if I know less and less. Life moves faster and faster. Your class motto reads, "People say the easiest part of life is over and the hardest is yet to come." We say, "Bring it on!"

I paused a moment. And then I turned to them, and I said,

"Oh, it's a-comin'."

I admit, their faces remained blank. My wisdom, wasted on the young. But out there in the bleachers, filled as they were with the over-forty crowd, came the wry rumble of a group chuckle. Message being, Class of 2003, may the wind be at your back, but as you hoist your clean white sails, take a moment to batten the hatches. I was grateful they were on the dais and I was on the gym floor, meaning when I delivered the line I was looking up to them, which is as it should be, because they are launching, while I am just paddling to stay in place.

○ ○ ○ ○

Apart from the tomatoes and lettuce, I still haven't gone full-tilt on the garden, but I count among the signs of spring the fact that the carp are running and the spirea are in bloom. When I was a child, spirea bushes ringed the front porch of our old farmhouse like a tall lace collar guarding a spinster's honor. When I moved into my current house, I was delighted to step out one May morning and discover the green bush outside my dining room window had blown up into clumps of white bloom. At first sniff I knew it was the same flower because I was immediately transported to that porch, where I would lie on a steel-frame bed and read cowboy books and the musty scent would roll through the screens on the back of a soft breeze.

It has become a tradition with me, this sniffing of the spirea. Once every May for nine years now. You have to pay attention, and then inhabit the moment fully, because the little florets do not remain long, shattering and falling like someone has emptied their miniature paper punch over the lawn. If it rains, you can lose them all overnight. This year I stand in the yard a little extra long. My life is full of great good, and gratitude remains at the top of my list, but there are too many appointments, too many deadlines, too many unanswered e-mails, too many *too many things,* and I could use some leaning out. For two weeks now the note on my computer calendar has read, *"Go through desk pile."* The spirea is telling me once more around the track and a decade will have passed, and maybe it's time to recalculate. I was upstairs writing today, trying to finish a piece for an editor waiting at a desk out east, when I noticed the blooms. I had moved across the room to open a window and let the warm air and noise of the day in, and caught the scent. When I went downstairs I felt the editor check his watch. Perhaps to prolong the moment, before I came back inside I picked one of the clusters and took it to the desk with me.

I have a reader's magnifying glass that I found in a box of things from an old man's house. The lens is rectangular and the handle is set at an angle to keep your hand from obstructing the page, and when I sat down I pulled it out to study the spirea.

The cluster is formed by a convex half-sphere of eighteen blossoms, each blossom balanced on a pale green stem the thickness of fishing line, each flower consisting of five overlapping white petals round and fine as flakes of detergent. You can see where the flower burst from the bud, the remnants of the sepals curled back, the calyx collaring the bloom like a soft green starfish. At the center of each bloom, the lobular carpels curl inward, forming a scalloped chalice with the waxy fullness and shine of squash skin. It is a fairy goblet, and from the very center of it all the stamens foist themselves outward in a spray of filaments, each knotted with a brown anther the size of a dust mote. It is good to see something this delicate. The scent, for all its evocative power, is not classically pleasant. For years I have struggled to put my finger on it, and I try again today. I hold the cluster to my nose and breathe in. Something dusty. The smell at the back of scratchy old curtains. The spirea seems always tied to the idea of windows, of the year's new air through the screen.

○ ○ ○ ○

Now I find myself in the waning hours of the final day of May, and as if I needed further evidence that I may be testing the limits of my calibration, I am in Los Angeles digging at the same cheeseball as the porn star Traci Lords. We occasionally work for the same company, although rarely in the same room; in this instance we are at a ginormous book fair recuperating after a hard day's work comporting ourselves as authors. I intended this to be the year of the truck and the garden. It has turned into the year of the truck and the garden when I can get to them, with a cameo appearance by an adult film star. Occasionally you get off track.

I revisit my gratitude list often. I have my health. I crank out my articles, I haul my books around in my Chevy, sometimes I get to fly while someone else hauls the books for me. I am spending more quality time with my truck and brother-in-law than I have in years. While my cousins and a few of my neighbors are abroad and in the line of fire, I am free to root around in my backyard without fear, unless the neighbors are flipping cars. There is a beautiful woman a ways south of here who will not bake my oatmeal bars but sometimes wears my flannel shirt. It's all good, as the kids say. But sometimes I think about those June bugs, zipping around apparently all fat and happy, but operating at speeds exceeding specifications, and this is turning into the year of looking for a way to bring it all in for a smooth landing.

I'll tell you what helps: I fly out of Los Angeles International Airport Sunday afternoon. Three hours on the plane and I'm in Minneapolis. Two hours in the car and I'm in New Auburn, watering my pig manure tomatoes in the twilight. Ten minutes and the hose is still in hand when the fire department pager goes off. Ambulance run in the county. A scared little boy who fell out of a moving truck. We bandage his scrapes and calm his mom. We are on a bridge spanning the Chetek River, and I can feel it moving smooth down there in the darkness. When we head back for home, the siren has blown L.A. right out of my dwindling hair.

Chapter 7

JUNE

THE SQUIRREL wars are on.

I am largely impotent in this battle, as the village has ordinances expressly forbidding gunfire (although the neighbor out back was once hauled off to jail for firing his black powder rifle up the alley—as the deputies stuffed him in the squad car he could be heard protesting it was only practice, and after all, he hadn't put a ball in), and furthermore my neighbor Charlie regards the squirrels as pets. Charlie saw horrific action in World War II, then came home to put in a half century of break-back farming before retiring to town, so if he wants to feed the squirrels, I'm going to let him. Still, a guy casts a gimlet eye.

As with most intractable conflicts, this goes way back. The first time I ever took a rifle to the woods, it was to hunt squirrels. They're tricky little buggers, especially if you hunt in pine forests, where they can so easily hide in the evergreen boughs. The trick is to run them up something deciduous—a leafless oak or maple. They'll flatten against the gray bark, become almost invisible, but you learn to look for a fluff of tail, the fine hair of which tends to catch the sun and move at the slightest breeze. You learn tricks, too. Squirrels almost always use the tree as a shield. Check it some time: Tree a squirrel, then walk in a slow circle in a radius fifteen yards or so from the trunk. Nine times out of ten the squirrel will edge around the trunk in skittery little increments, always keeping the trunk between him and you. Sadly for the squirrel, he can be reliably fooled. If I was hunting with my father or brothers, we would approach the tree together. Then one of us would stand stock-still while the other walked the circle. Focused on the moving person and apparently unable to do math, the squirrel would edge around until he was exposed to the frozen guy with the gun. The statue technique works if you're alone, as well. Once the squirrel is treed, you lean against a tree of your own, and just wait. It may take ten minutes, but eventually the poor fellow sneaks a peek, and you've got your shot.

Yessir, I despise the squirrels, but I haven't killed one in years. Mainly because I am old-school on the eat-what-you-shoot rule, and I've never put squirrel high on the tasty list. Some around here call them tree rats, and although well-meaning nature-meat lovers will try to convince you otherwise, they taste that way. Rubber chicken, my dad used to say. Unlikable, too. Anyone who has ever sought solace in the deep wood only to have a squirrel perch three branches away and deliver a thirty-minute scold knows they are nature's equivalent of a nosy neighbor.

But what gets me going in June is they deprecate my garden.

The morning began poetically enough. I set about transplanting cherry tomatoes and unearthed a logy June bug, which, as it slowly clawed the air, reminded me of the dung beetle in the opening paragraph of Franz Kafka's *Metamorphosis*. So I was feeling self-sufficient and well-read. Although I figured I had done in the June bug. He was metamorphosed enough to wiggle his legs, but the fact that he was still underground suggests he wasn't quite perfected. He is probably due for a robin, or a merciless grackle.

When I finished with the tomatoes, I fetched a tray of basil sprouts and one of cilantro and transplanted the works. I retired to the house with the usual sunny feelings imparted by working bare-handed in the earth. Ten minutes later, when I passed the raised bed on my way to fetch the mail, the basil and cilantro were uprooted and strewn to the six corners of the garden. There were telltale claw marks and prospect holes the diameter of walnuts. Balefully I scanned the trees.

I was able to tuck most of the plants back in, but this speaks to why I so despise the squirrels. They don't even *eat* the stuff. They just rip it up in some misguided fit of excavation triggered by oh, I don't know . . . the smell of *dirt*? At least a rabbit ingests a sprig or two. And rabbits *taste good*. Squirrels are vandalistic *and* inedible.

And the *swiftness* of it. After several years of this, I have taken to stapling nylon netting over the beds until the plants have rooted and established themselves to the point that the squirrels don't usually bother, but I was hoping to do a little more planting today, and figured for the love of Pete that I had time to walk half a block to the post office. I should have known better. There have been times in the past when the squirrels have blasted through a row of leeks in the time it took me to run to the basement for a packet of pumpkin seed. Do they think the big dumb human buried a nut? And now here comes Charlie with his corncobs. As if the chattering herds of obese sky pigs that rule my lawn aren't already the size of otters. And the thing is, they drag half that corn over here and bury it in my grass. I can't grow a bloody thing in my garden, but I've got corn sprouting every two square inches. It looks like the test plots of the University of Wisconsin Corn Agronomy Program out there. I don't so much mow the lawn as make silage.

What we have here is not a backyard but a rodent preserve.

○ ○ ○ ○

Regarding the state of poetry, I am willing to yield the balance of my time to Camille Paglia, but from the grandstand of Red Cedar Speedway in Menomonie, Wisconsin, it appears that if written verse is to recapture its relevance in popular discourse and contribute to the promotion of universal human understanding in any contemporary sense, we must convince more poets to attend dirt-track stock car races.

Heaps of grant money will be required.

In fairness, it is probably easier to get a poet to attend a stock car race than it is to convince a stock car fan to attend a poetry reading, although I once scanned the audience midway through a reading and recognized the winner of the previous weekend's Open-Wheel Feature, an experience analogous to that of the birder who spots a pelican at the hummingbird feeder. I invoke the term *dirt track* with the specific intent of distinguishing down-home fender-banging from the NASCAR circuit, which is rapidly achieving a level of manufactured drama just a hair south of professional wrestling. Even more disturbingly, NASCAR references have begun popping up in *New Yorker* cartoons. In a yin to the yang of the Wal-Mart yoga mat, the hallowed sport of Carolina bootleggers has evolved into a corporation of sound-biting action figures with beautiful teeth. The speed and danger are as real as ever, but the carnival is fueled as much by merchandise and central casting as 110 Leaded Racing Gasoline. I prefer the local dirt track.

I wind up at the dirt-track races a couple of times a year. Not enough to claim Number One Fan status, or recite the standings, or even provide a clear description of the various divisions, but enough to talk strategy and revel in the evening. Tonight when my friend Gene and I step out of the car in the parking lot behind the grandstand, the first thing we hear is the roar of the racers running warm-up laps. Then the prevailing west wind carries the caramelized scent of combusted racing fuel to our noses, and now we are hurrying a little bit, eager to get inside the gate to see the cars and all the people and fold into the scene. I hold my binoculars and earplugs in one hand while fishing out my wallet with the other, and then the woman in the booth at the gate takes my money and slides me a ticket. "Keep the stub, hon," she says. "There'll be door prizes."

As we pass through the main gate another woman takes and tears our tickets, handing one half back and stuffing the other in the door prize pail. We make our way through the churning crowd past the restrooms and the food concessions and pause for a minute at the T-shirt tables set up just outside the grandstand entry. Even at the amateur level, racers understand the value of merchandising. Mostly it runs to clothing decorated with the racer's number and sponsor logos (SHIRTS $15.00; SWEATSHIRTS $20.00; JACKETS $70.00), but you can also get window decals (SMALL $4.00; LARGE $5.00) and many of the racers sell die-cast models of their cars, complete with graphics and lettering. Sales are usually handled by second cousins and girlfriends. We sidle past the arms-folded sheriff's deputies, and now we are trackside, on the dirt path that runs between the stands and the retaining wall with its chain-link fence, cabled in theory to keep the cars off your lap, a slim reassurance when they blast past six feet away, so close the flung dirt clods sting your neck.

The people in the steel seats take all forms, but mostly they run to overbellied men, couples in matching bar jackets, tots with their heads clamped in oversized hearing protectors, and, above all, many replications of a certain kind of young man best summarized as a conflation of frat boy and redneck, with the affectations of both: goatee, maybe an earring, ball cap, well-executed tattoos, and a cud in the lip. The ball cap will be worn straightforward and the bill will be worked into a tight curl. These boys drive shiny four-wheel-drive pickups, often with a four-wheeler ATV parked in the box. They are not big on walking. When they do get caught afoot, they carry themselves with a blue-collar insouciance, but there is an overfed softness to their profile that belies all the time spent sitting atop four wheels. In the stands at the stock car races, their natural stance seems to be jacket open, belly out, one hand down the front pocket of their jeans, the other hand wrapped around a beer.

Contrapuntal to these boys is an archetypical sort of woman, slim-hipped to the point of boyishness and wearing jeans approaching the melting point of tensile resistance. Her hair is harshly blond, long in the back and shellacked to a pouf up front—the feminine mullet, or *femullete*. She is likely wearing a team T-shirt, white, with her favorite race car rendered in garish fluorescents on the front, and the season schedule on the back. Women like these have a little edge to their eyes, it's part life and part cigarettes, but when you grow up where I did, you imprint on their type so that when I see them down front there, walking in pairs to the restroom, they always trigger a reflex attraction. Something atavistic. If you fancy yourself a lad, you should keep in mind that most of these women own a pickup and a deer rifle, and odds are they can take a punch. Women like this, you make a wrong turn, and they'll put you in the ditch.

There is this moment, when they're getting the tape of the national anthem cued up, and the shiny vehicles from the local car dealer are parading around the track with the flag, and the local princess of something is waving from a convertible, when it gets so quiet you can hear the tires padding over the clay. Then the anthem kicks in and we all stand, and the combination of patriotism, anticipation, and open air makes my chest swell and my guts tingle, and it seems that what you have here is a moment of unification, a convening of like-minded people joined by dirt, noise, and country, and you are unapologetically thrilled to be part of it. It is, I think, the absence of pretension that frees up your soul.

Then they line'em up and cut'er loose. It is a delightful maelstrom. The cars, all sheet metal and pipes and bellowing speed. The track, reddish and soft early on, growing rock-hard, rutted, and iridescent with rubber as the night wears on. The feel of the grit that settles on your skin in the wake of every pack. And always in the air that scorched cotton-candy grace note of spent racing fuel. The spectators pay close attention to the action, cheering their favorites, doing their best to hex the bad guys. Gene, by day a mild-mannered Birkenstocks-wearing physical therapist and owner of one Volkswagen and one Volvo, is pumping his fist and whooping it up like Junior Johnson's redheaded stepchild.

○ ○ ○ ○

Harley Paulsrud is widely considered to be one of the bad guys. You'll see plenty of his T-shirts in the stands, but an equal number of people will cheer up a frenzy if he tags the wall. Back in our glory days, Harley and I spent every autumn Friday night on some football field, home or away. I played left defensive end. Harley played left outside linebacker. It was up to us to stuff the option. I'd turn the quarterback up, force the pitch, and Harley would tee off on the running back.

Harley was a hothead, always ready to fight, always ready to hit, on the field or off. But he was also one of those rare athletes—John McEnroe comes to mind—who seemed to be able to turn anger to his advantage. Rather than become rattled, Harley seemed to hone his anger like a blue flame. We had a mediocre season our senior year—3–3 in conference play, as I recall—but Harley and I had some fun. We got to where we keyed off each other, knew what the other would be doing without looking. By then Harley and I went back thirteen years, clear to kindergarten. I was the oldest child in my family, he was the youngest in his. I remember him drawing a picture with his crayons that first year, a green helicopter over a field of fire. I didn't know it then, but his oldest brother, having survived a tour in Vietnam, had just died in a helicopter crash on a base in Texas. This only two years after Harley's father was crushed and killed when his tractor pitched over on a hillside. You see, maybe, where Harley came by some of that anger. In music class one day, first or second grade, I accidentally whacked him on the shin with a music stand. He flashed red with rage. "You just wait until recess," he hissed. I was scared, but determined not to show it. "I'm waiting," I said coolly, my knees quaking. He shot me nasty faces all the while Mrs. Carlson trilled away at her piano, and he glared at me all the way back to the classroom. Every time I thought of recess a cold finger tickled my liver. When it came, all the boys boiled out of the doors to choose up sides for football. Usually I would play, but today I hung back, on the sidelines over by a snowbank. Harley got the ball immediately, and headed right for me. He slipped on the ice, but as he slid toward me, his feet were windmilling and he managed to kick me two or three times. And that was it. It was over. It hadn't really hurt, and I felt a surge of relief. Looking back at it now, thinking about how I faked a cool pose in the music room, how I slogged out to recess even in my dread, I realize Harley put me through a critical toughening process. He taught me to act like I wasn't scared, even when I was. Every time I swallow my fear and move forward—toward a fire, a bad wreck, another disillusioned girlfriend, the dentist—I am exercising a conditioned response traceable in part to Harley Paulsrud.

Harley and I pretty much lost touch after graduation. I heard he was working highway construction and I'd see his name in the racing reports published by the local weekly. Now, twenty years after our last tackle, I am pointing out his blue and gold #1 car to Gene. Harley races in the WISSOTA modified class—essentially, he is driving an open-wheeled stock car. The races go on, class after class, heat after heat, and during Harley's races, I follow him intently. I can't see his face, so I focus on his hand, clad in a blue glove and visible at the wheel. If they stripped the number from his car, you could still tell it was him, just by the kamikaze way he attacks the turns, and the way he dives low to make a pass, or shoots up and takes the high line on a blur, flashing through the grandstand straightaway inches from the concrete retaining wall.

I like to think of the tiny sounds in the midst of all that pandemonium: the creak of the seat belt, the zap of the spark plug arc, the smooth spin of a bearing. Likewise, all the hurtling steel is being directed by the tiniest of movements: a quarter-inch tug on the steering wheel transmits somehow down through the shafts and pinions, the shuddering springs and the spinning wheel, right down to the clay, where the rubber adjusts its purchase and the car jukes or straightens accordingly. The modified is a powerful class of race car. Stuff the accelerator too deep too quick and the tires will "lighten," spinning at the sacrifice of traction, at which point you hit the wall like a can in a crusher. Drive with an egg under your shoe, the old-timers say. At the heart of the vortex, the cars are positioned according to the sum total of these invisible nuances, but from where Gene and I are sitting, it is simply a stinky, thunderous, and cathartic chariot race. Sense-wise, it doesn't hold up under examination, but then neither does two-man luge.

Even a stranger would pick up on the fact that Harley has the crowd evenly divided. When he makes a huge pass and moves up three places, half the crowd cheers. When he gets flagged for an infraction and sent to the back of the pack on a restart, the other half cheers. Just before the flag drops on the restart, I lean over to Gene and say, "Back when we played football, the angrier he got, the better he played." And sure enough, when the man in the crow's nest waves green, Harley drives with a vicious focus. You can feel him seething in the way his car snaps and darts from turn to turn. Starting dead last, he picks off car after car. When the checkered flag drops, he is up to third place, with first and second well aware that they owe their places to an expired lap count. Another few turns and he'd have had them, too. When Harley comes down the final straight, I jump up and holler him across the line. Gene, who can teach you to walk after you get hit by a truck but requires six hours and two sets of backup filters to change the oil in his Jetta, is wide-eyed and grinning. He pulls his earplugs.

"Man, once he got clear of the dirty air, he just *flew!*" *Dirty air* is racing slang for the turbulent wake left by the car ahead of yours. I fix him with a look.

"Gene, have you been watching *RPM Tonight* again?" He ducks his head and blushes.

I can't blame him. You get swept up. I get these calls: "Mike! Friday night! Punky Manor Challenge of Champions! Hot racing action!"

Keep it up, I tell him. They'll yank your physical therapy license *and* your Sierra Club card.

In the feature—basically, your nightly championship race—somebody tags Harley from behind, kicks him sideways in Turn Four, and then the cars pound him, seven in all by the time corner marshals get it all shut down. The cars are stacked and crumpled together, and it takes awhile for the tow trucks to sort the mess. I can see Harley in his roll cage, trying to start the engine, and then I can see a red light come on, which I suppose is bad news, and eventually they tow him back behind the wall, and I imagine he threw some wrenches. Still, he is headed for a good year, with enough points to give him a shot at the state championship.

I was surprised at my reaction, seeing Harley out there. It was exhilarating to watch the figure in the screaming sheet metal knowing we had history. To look at that hand on the wheel and remember how many times we had knocked some quarterback flat and then grabbed hands to help each other up. To know we could fill a night with stories and just be getting started. When I got home I looked up his address in the phone book and sent him a post card, just a simple note telling him how much I enjoyed his performance.

◇ ◇ ◇ ◇

Revenge fantasies are always a sign of emotional delamination, so perhaps it is unhealthy that lately I have been wondering just how difficult it would be to rig the raised beds with antisquirrel land mines. An acorn's worth of C4 should get the job done. This time they reached right through the nylon netting and fiddled out several leaf lettuce plants, which were just getting to the picking point. I cussed when I saw this, cussed out loud. Had he not been rendered stone deaf by World War II ordnance and thirty years on a Massey-Ferguson combine, Charlie might have heard me, because he was out there with his corn again, re-stocking his squirrel feeders.

And this is where I'm not entirely clear on whether Charlie loves the squirrels or not. True enough: whereas I would quite gladly honey-comb the yard with squirrel-sized punji pits and perch in the walnut tree with a blowgun, Charlie faithfully lugs enough corn into his back-yard to feed a half-dozen beefalo. But he doesn't just turn it over to them. Charlie is a fan of what I call trick feeders. Essentially dribble glasses for the backyard set, trick feeders force the squirrel to endure a variety of humiliations in order to get a taste of corn. Your quintessential product would be the Squngee, a perverse contraption that dangles two cobs of corn from a bungee cord. While the squirrel fights to shuck and pocket the kernels, the cob bounces up and down. Some models come with a bell attached. Lately the Squngee has been mass-produced by prison inmates.

Charlie has a trio of trick feeders. The mildest of these is the minia-ture chair feeder, in which a corncob is skewered on a nail before a tiny chair. In order to get at the cob, the squirrel plants his little hinder on the seat. Variations on this theme include miniature Adirondack chairs and picnic tables. Nailed to the tree beside the miniature chair, Charlie has another device that, at first glance, looks like your standard bird feeder—a little roof and deck—except that the corn is stashed in a glass mason jar screwed to the front of the feeder. In order to get to the corn, the squirrel has to crawl in through an aperture on the side of the feeder and into the jar, where hilarious contortions ensue—all visible through the glass. Finally, Charlie has rigged up a bicycle wheel with spikes so that he can stud the rim with cobs in the manner of a ship's wheel. When the squirrel ventures out and grabs a cob, his weight sets the whole works spinning, and as the squirrel fights to hang on and eat, you just laugh and laugh, which makes it tough to get the little bugger in the crosshairs.

Anneliese and Amy are visiting for the day. Amy is sitting beside me on the edge of the raised bed as I pull weeds. I like to pull weeds in the morning, especially if I remember to water the night before, as the weeds come up easier. Pulling lamb's quarters is most satisfying. The stem is strong and you can get a good purchase. The roots come out in a neat clod, as opposed to quack grass, which tends to snap off. There's a nice rhythm to weeding when you really get going, pick-pick-pick, pull-pull-pull, the miniature clear-cut widening as you proceed, the vegetable sprouts seeming to gather strength and shift up a gear almost immediately upon the removal of the competing roots. And then there is that calm satisfaction you feel when you look over the freshly weeded bed, everything clean and neat and bound to flourish.

Amy pulls a few weeds, then wanders off to play. At the moment, the best she has been able to come up with toy-wise in my house is a hand-carved, gape-mouthed wooden hippopotamus from Africa and a firefighter doll I was given as a thank-you for speaking to a group of local schoolkids. She brings me the hippo and informs me with some gravity that his teeth are loose. I turn the firefighter into a dentist, and he tightens the hippo's teeth with an impact wrench, or at least that's the sound I make. Amy produces a candleholder. The firefighter uses his hose to fill the holder with imaginary water, after which the hippo swishes and spits, and then they are all off on some other adventure.

I weed a while longer, then switch to planting. As expected, the nasty cold of last winter, combined with little or no snow cover and my failure to mulch, killed off most of my perennials. I started some sage in the basement, and after transplanting that, I plant rosemary and lemon balm. It's nice to chop up the dirt with the hand cultivator, smooth it with my palm, and then draw straight lines with my fingertip. I drop the lemon balm seeds in, remembering how the leaves taste plucked and eaten raw, or torn into chiffonade for vinaigrette. I can see Anneliese, back from a run, moving around in the kitchen. The sound of my cheap refrigerator radio seeps through the screen.

I go back to weeding, and Amy returns again, this time with a question about what the hippopotamus should eat, but then she sits in my lap and starts pulling weeds herself. As we work I feed her sprigs of lettuce, and we make jokes about squirrels stealing my salad. For a good twenty minutes, I weed and she jabbers, her little blond head right there beneath my chin. At one point she looks up and says, "You be Amy. I'll be Mike."

I am falling in love twice here.

○ ○ ○ ○

Way up in northern Wisconsin, in Bayfield County where the summer
sailors go, in a clearing on the slopes of Mount Ashwabay with a view to
the Apostle Islands of Lake Superior, stands a bright blue tent capable
of holding nine hundred people in a way that makes them feel they are a
cozy dozen. This is Big Top Chautauqua, an institution born of one man's
love for music and canvas. Officially classified as a "non-profit performing arts theater," the Lake Superior Big Top was first pitched by Warren
Nelson in 1986 and in its current incarnation (the tents wear out, and one
was lost to fire in June 2000) opens its stage mid-June through early September for musical comedy and historical revues (including the recent
angling-based smash *Riverpants,* written by Warren and including giant
projected photos of Nice Fish), family matinees, and a steady stream of
national folk and country acts. Johnny and June Carter Cash have played
here, Willie Nelson, Emmylou Harris, John Prine, Leon Redbone, Earl
Scruggs, Judy Collins, Bruce Cockburn, BeauSoleil, the list goes on—
some twenty years long now. Anneliese and I are here to see the singer
Greg Brown. We are on our first weekend away together. Amy is with
Anneliese's mother. We will act grown-up and dally at will. Yesterday
Anneliese ran a half-marathon in Duluth while I read the newspaper
and ate a danish. The night previous we slept in a tent on the outskirts of
Superior, Wisconsin, and it was cold, cold, cold, but among other things,
love is Sterno. Today we lazed the Chevy west along the southern shore
of Gordon Lightfoot's Gitche Gumee, following Highway 13 past Port
Wing and Cornucopia (stopping to buy a used book about Lenny Bruce,
and a packet of smoked fish from a pretty woman in yellow rubber overalls) to Red Cliff, then finally hung a right and dropped six miles south to
a plain motel, where we took a room overlooking the water.

During the drive we talked about how it's going with us, and we
are daring to hope. Just about three months in, and we feel a certainty.
Among other things, we would both like a small farm with a barn and
chickens. We are still operating within that window when joy and discov-
ery reign, when it is very purely enough to be driving and holding hands,
but we have both lived enough to understand this and enjoy it anyway.

Or especially. I have failed at points far, far beyond this, but I have never looked into another pair of eyes and felt such ease.

Greg Brown is a hard lesson for well-groomed boys everywhere. Snaggle-toothed and beefy and mumbly and prone to performing in clogs and overalls and not looking up much, he appears careless until he sings, and naturally this makes him irresistible. *I'm a Midwest boy*, he sings, *I'm a big dumb man . . .* and you can hear the murmur as women from Iowa City to Manhattan give the ol' thumbs-up. In the heartland, Eros tends to clomp around. Lesson being, tweak and trim if you will, cover boys, but don't overwhelm manhood with maintenance. There is nothing *citrusy* about Greg Brown. The poetry and guitar help, of course. Quatrains make bad news beautiful, and any given six-string will smoke-screen a raft of shortcomings.

Above all else, there is that voice. Sometimes I drive all night drinking truckstop coffee and singing radio notes exceeding my range at both ends, or I holler-talk elevated nonsense through the secondhand smoke of some bar to 2 A.M., and the next day when I rise my larynx is so scorched and seasoned I can make my sternum buzz in and out of phase just by humming, and I'll think, this is what it's like to possess the vocal cords of Greg Brown, but even in my most sonorous state, I am pleather to his leather. Greg Brown's voice sounds as if it was aged in a whiskey cask, cured in an Ozarks smokehouse, dropped down a stone well, pulled out damp, and kept moist in the palm of a wicked woman's hand. I think if he says good morning across his coffee cup, it raises ripples. The voice is a perfect match to his lyrics, biblical and bar stool and garden loamy as they are, all Rexrothian and as easy-rolling true as a brand-new '64 Dodge. A Greg Brown song doesn't make me want to whoop and holler, it makes me want to sift bare-handed through the dirt for repentance and then go looking for a woman who doesn't mind a few chickens. I was hoping Anneliese would like him.

We drive to the Big Top early, so we have time to sit and watch the people gather. Ski Hill Road originates at Highway 13, breaking westward on the perpendicular to proceed in a gentle mile-and-a-half climb through close trees to the ski hill gate, where we present our tickets and

park in the gravel lot. After buying coffee, we perch on a berm near the
end of the T-bar tow. I lean back against a wooden pole and Anneliese
leans back against my chest. The coffee tastes startling good against the
clean air. The sun is easing lower, pushing its gold through the dust haze
raised by the arriving cars. The tent is a blue burst against the surrounding green, looking from a distance like a squat storybook caterpillar with
stripes of pearl gray. There is an eager civility to the crowd, everybody
milling but quietly so, folks passing in and out of the tent and clusters of
friends meeting up to have beers in the grass or eat brats in the conces-
sion tent. The trees are optimistically green. I put the coffee down and
link my hands across Anneliese's lap, put my cheek beside her cheek.
The air is soft and warm, and from somewhere comes a whiff of patchouli with its attendant evocations of wooden-floored head shops and
lace-up maiden dresses.

There are certain givens at a Greg Brown concert, patchouli among
them. The parking lot will be populated by a high proportion of Subaru
wagons, the number one automotive accessory being a dead-heat tie between kayak rack and dream catcher. Regardless of make, choose any
vehicle at random and a minimum of two radio preset buttons will get
you National Public Radio. In the backseats and cargo areas you will spot
fishing poles and Frisbees, and here and there a rain stick. Many of the
passengers exit the cars carrying rainbow falsa blankets and Guatemalan
tote bags. The men tend to wear hiking boots or leather sandals and
floppy-brimmed hats purchased by mail order. A lot of the women wear
wrap skirts in Indian print, crepe, calico, and tie-dye. And to be fair, a lot
of folks are dressed like weekending insurance agents, after-hours physi-
cians, and vacationing teachers. Jeans and fleece pullovers, deck shoes
and hikers. But by and large, the crowd has a certain well-fed alternative
look. You could not go far wrong casting the Greg Brown audience as
mainly hippies with health insurance.

I'm of the leather sandals persuasion tonight, having left my usual
steel-toed boots in the motel. Baggy shorts, no socks, a heavy shapeless
sweater in case it gets cool. Fashion is hopeless with me. Don't have
the body for it, for one thing. The runways of Paris and Milan are rarely
devoted to the short and stocky among us. I flailed around after fashion

until I wound up on the road with a pack of country music roadies some-time in the mid-1990s, and taken by their functional and durable ap-proach (heavy boots; long, baggy shorts with pockets; free black T-shirt), I adopted the ensemble and have never really progressed. Anneliese is wearing a gypsy skirt. Her long blond hair is parted simply and falls to either side of her eyes: blue, so blue, and one with a fleck of brown, which seems to symbolize luck.

The light fades from the hills and the crowd gathers and turns itself inward, as if the people were metal filings and the big tent—lit now from within by bare bulbs strung through the quarter poles—were an elec-tromagnet on a rheostat, with Warren Nelson slowly juicing the power until everyone pulls into lines and is drawn to the center. Anneliese and I have seats fairly far to the rear, but we have a clear view of the stage. At Big Top, every seat is a good seat. On a night as warm as this, Warren rolls the tent walls up beneath the scalloped canvas eaves and spillover audience members watch from camp chairs on the lawn. When Greg Brown walks out to his chair at center stage, a big cheer goes up. Some guy seated way in the back yells *"WOOOO!"* but it is a decorous sort of hooting along the lines of *Remember that time in '92 when we got ripped on microbrew and totally wrecked the recumbent tandem?!?* Brown ac-knowledges it all with a bump of the head, then gets to work. No laser lights or flash pots, he just sits in a chair and begins to play.

It could be the rosy cusp of love, but it's a wonderful show. Anneliese and I listen hand in hand, leaning together, letting the easy evening move through us, the silhouette sea of heads before us, and Greg up there on his chair singing that he loves it when his baby calls his name, at which point Anneliese looks up at me and smiles and I—projecting the way all men do when it's another man up there singing—grin, thinking, *I'm a Midwest boy, I'm a big dumb man* . . .

○ ○ ○ ○

When we get back home, I find the squirrels have had another go at the cilantro. Some of it has lain there a bit, and will not recover. I reroot the ones that still have traces of starch in the stem. I am rethinking my Eat No Squirrel policy.

Charlie is on the fade. He has always been a thin man—I retain a vision
of him when he came to harvest our oats, standing atop that Massey Ferguson combine, his jacket hanging on his high wide shoulders like folded
vulture wings, his gaunt face and sharp nose completing the image. Now
he's eighty-six years old and when we talk in the backyard and he draws
in close so he can hear, I notice his nose is thin as a turkey breastbone.
He always looks cold and pale, even on sunny days, and he wears his
jacket and a billed cap. Once we were talking and he suddenly sagged,
sinking to one knee. I locked up for a second there, debating whether I
should grab him or run for the phone, and then he simply stood back up
and started in talking where he left off. Oftentimes he asks me how the
farming is going, and I realize he is confusing me with my brother Jed,
the true farmer among us. But he seems content, filling his feeders and
sitting for hours cross-legged on his chair on the little deck, leaning over
every thirty seconds or so to squeeze loose a viscous string of snoose.
Now and then his wife, Toots, loads him in the car and they head up to
the casino to play some bingo and have the buffet.

Using directions I found on the Web, Mark has been rewiring the International from its orginal six-volt system to twelve volts (you note the
division of labor). Last Sunday he called to say he had changed the oil
and filter and thinks we can try to start it. Now it's Friday night and Anneliese, Amy, and I have driven over for dinner. Mark has removed the
oil bath air filter and he leans in from the side to prime the carburetor. It
has been years since I started the truck, so when I climb into the driver's
seat and lean forward and reach to the right to twist the key and to the
left to punch the starter button, the geometry of my body releases a flood
of memory, and I am imagining what it will be like to go blasting down
the Swamp Road again. The truck is still missing its bed and the wheelless front end is propped on jackstands, but still. Mark dribbles gasoline
into the carburetor and nods at me through the windshield. I twist the
key and punch the silver starter button, and right off the bat she fires,
grabbing for just a couple of spins before the dribbled gas is used up,
but oh, our goofball grins through the cracked windshield in the silence

after the engine kills. O, glorious noise! What a machine, to sit there neglected, then leap to life so easily.

Nothing is so simple, of course, and the problem is that the fuel pump seems to be failing. As long as Mark dumps gas in through the top of the carburetor, the engine will run, but there is a clear glass bulb on the carburetor, and as the engine spins, it should be working the fuel pump, which in turn should be pushing gas up a copper line and into the glass bulb with its float. But we spin the engine for a good long while, and no gas appears. Short of duct-taping Mark to the fender so he can trickle the gas while I make dump runs, we're going to have to sort out the problem. This is where even a poor mechanic like me has a chance to contribute, because the elements are fairly simple, and you're working through a process of elimination based on equal parts mechanical knowledge and conjecture. It's a form of algebra. Using known quantities, you're solving for the missing part.

There is a danger to working problems like these. You tend to get caught up. When I was still in college and driving the truck full-time, I had a problem with the engine stalling out at high speed. High speed being a relative term, of course, as we are talking in this case just shy of 55 miles per hour. The truck would start fine, it would run fine, it would go through the gears fine, and it never had this problem in the city, but at some point after I settled into the high-gear whine on the four lane, she'd hiccup, then hiccup again, and then the power would just go out of her. I could coax out a couple extra tenths of a mile by fiddling with the choke, but eventually you'd have to back the accelerator all the way off, coast awhile, and start over. I was flummoxed from the get-go, and it took my brother John—fiddling on the truck while I was back at college—to figure out the problem, which in the end was fundamentally simple: a small fragment of twig just slightly larger in diameter than the fuel line had fallen into the gas tank. When you were moseying around in the lower gears or at low rpm in high gear, the twig just bobbed around harmlessly. But when you got to pourin' the cobs to her for a long stretch and really sucking the gas, the twig would eventually float around and be drawn into the fuel line, where it would obstruct the flow. When you let off the gas, the suction diminished, and the twig floated back out and

you could hammer until it got sucked back in. I had forgotten about
this problem of years ago but it all came back tonight when I saw Mark
leaning in to pour gas in the carburetor. At one point, when John and I
were trying to figure why the engine was starving for fuel, we decided
the only way to tell if the bottleneck was occurring between the tank and
the carburetor was to run some tests under live operating conditions. At
the time, we were up the road at Jed's farm. I removed the hood and the
air filter and got in the driver's seat. Then John spread-eagled himself
over the engine compartment, wedging one boot against the firewall and
one against the battery rack. He grabbed a hood latch bracket with his
one free hand. In the other he held an aerosol can of highly flammable
starting fluid.

"A'right," he said.

I pulled out on to Beaver Creek Road. We had a straight shot all the
way to Baalrud's Corner, which is just shy of one mile. By the time we
got to Mom and Dad's place, we had a pretty good head of steam, and
by the time we passed the last of the buildings we were flat pickin'em up
and puttin'em down, nothing left to give. John had a death grip on the
latch bracket, and his hair was lying flat. His beard was wrapping back
around his neck.

Upon reflection, you just don't know what you were thinking. I mean,
me, sure, I've got a proven record as a bit of a ninny, but that brother of
mine, he's straight-arrow sensible. Been tut-tutting around like a seventy-
year-old man since second grade. I was always the one screwing around
on the school bus while he sat there prim as Becky Thatcher. But this is
why I love my brothers, because when it comes—as Louis L'Amour used
to have his cowboys say—to *cut'n'shoot*, to *root-hog or die*—my brothers
are good to go.

And sure enough, right about the time we crossed the west end of
North Road, right about the time I was thinking I might have to back
out of it so we didn't overshoot Baalruud's Corner, she cut out. Bogged
right down. John brought the bottle throttle to bear and shot a snort
right down the old girl's throat. *Bwwaaarrr,* she came right back. Then
she faded, he gave her another snort, and *Bwaaarrr,* off she went. Pesky
automotive experts will frown on dousing your pistons with unadulter-

ated doses of rocket fuel, but the experiment served its purpose, narrowing the trouble down to the fuel line. It was based on this information that John hypothesized the stick in the tank. But him hung out there on that hood! If a deer had crossed, or some local had shot the stop sign on North Road, or if I'd blown one of those bald front tires, we would have had ourselves a gold-plated Darwin Award.

Years later I saw a television show featuring similar stunts performed by younger versions of my brother and me.

Jackass, they called it.

○ ○ ○ ○

The missing front tires prevent Mark and me from performing similar tests, and so we are reduced to head-scratching and mulling. First we pull out the old manual and study the fuel pump schematic. Mostly there is silence and foot-shuffling. Then we look the fuel line over with a trouble light. I find a drop of fuel up where the fuel hose attaches to the carburetor inlet, and although a leak is a bad sign, we take this to mean that the fuel line is priming, which is good news. It just hasn't reached the carb yet. Sitting as long as it has, the fuel pump diaphragm may have dried out and needed to wetten and expand before it could form an adequate seal and begin pumping effectively. I jump back in and punch the starter while Mark dribbles gas. We spin the engine this way for a long time, but still there is no fuel in the clear glass bowl. "Could be sucking air," I say. Mark tightens the hose clamp. He dribbles more gas, I start the engine, we watch the bulb: still no fuel. "Let me try something," Mark says. He goes around behind the cab to the fuel fill pipe. It has been taped over to keep dirt and dust from getting in. He pulls the tape. "We might be getting airlock," he says. You see how your head works with these problems. We spin the engine another good dose. Nothing. We go over the fuel line again. I take the trouble light and creeper down below, and find a loose fitting on the fuel pump. Up above, Mark finds a missing hose clamp and replaces it. This time, instead of dribbling gas, he fills the carburetor bowl full beforehand. I start the truck and come back around to the front, where together we stare at the bowl while the engine idles. The fuel slowly drains away, drawn into the engine, then the bowl empties and the engine dies. Mark fills it again, and we repeat the process. Again, the fuel drops to empty and the engine peters out. We run the fuel line one more time, but find nothing. Mark fills the bowl. I start the truck. Once again, our faces fall at the same rate as the tide receding from the square glass bowl. And then, just as the last of the fuel is disappearing, fresh gas comes spewing into the bowl, all amber and frothy like a poured beer, and the engine grabs and just keeps spinning, and spinning, and we jump back from the truck and slap high five. When we come in for dinner, Kathleen and Anneliese say they could hear us woo-hooing clear up in the house.

❁ ❁ ❁ ❁

Saturday. Anneliese and Amy have gone back home to Eau Claire, and I'm back at Mark's shop. Tomorrow I am flying to New York and won't be back for a week, so I want to put in a full day. From outside the shop, I can hear the buzz and hum of the welder. Mark is fabricating a diamond plate patch for the front edge of the truck bed, where the rust has chewed big holes in the steel.

I'm going to strip out all the hoses, including the heater hoses. The rubber is cracked and rotting and needs to be replaced. Then I'll pull the radiator. *Gonna hafta pull'er,* yep. We waited to do this until we knew the engine would run—last thing you want to do is replace all the hoses, then have to remove them all again to tear the engine down. I run the creeper under the front bumper and pull a work lamp in with me so I can see to loosen the petcock that drains the radiator. It comes open easily, which is a nice surprise, and the green fluid streams out in a fine arc and into a pail. The antifreeze is still running out when I loosen the pipe clamp on the lower radiator hose, and like an idiot I pull the big hose anyway, which means the remaining antifreeze flushes out all at once, over my hands and into the work lamp, which of course explodes and dies. My hands are slick, and I have to shoo the dog away from the green pool spreading across the floor. You can't spoil my spirits though, because this is my kind of work. Simple deconstruction.

After decades under the clamp, some of the rubber has congealed to the pipework, and I have to cut the hoses away with a utility knife and nippers. When they've all been detached, I start loosening the studs that hold the radiator in place. I drop one and it rolls off somewhere. When I can't find it, I make a note of it and continue. I get all but one stud loose. The last one snaps off and we might have been hung up, but Mark blows it out with the plasma cutter and the radiator is free. We lift it out, then lift out the shroud and cowling, and I put the radiator in the back of my car. There's a guy a few miles up the road from my house with a radiator repair business. I'll take it to him.

I work all day, get home real late, and still haven't packed for New York.

It's 2 A.M. before I make St. Paul, where I crash on a friend's couch after ordering a taxi to pick me up three hours from now. As I drift off I have a vision of Anneliese and Amy sleeping in their dark house so many miles from here. Before I get back from New York, they will fly to Colorado to visit Amy's father for two weeks. I will join them for the second week. I am wondering how Amy's father and I will circle each other. I am told he drives a Subaru wagon. I think about Harley and how he taught me to get out there for recess even if I am scared. We'll see how it goes.

In the trees above my garden the squirrels sleep, dreaming their evil squirrel dreams.

Chapter 8

JULY

IN JULY, EVEN the lamest garden gets rolling, and I should be at home weeding. But I am in New York on business (a euphemism meaning *keep your receipts*), taking meetings with editors and agents and assistants—strange shadowy figures who for the other 360 days of the year exist only in the form of e-mail addresses or voices on the telephone. During particularly lean junctures in my career, the thought has crossed my mind that these people don't exist at all, and someone in a cubicle somewhere is having a very good time with me. It is reassuring then, during my infrequent visits, to see them manifest, an added benefit being, when in physical form they pick up the check for lunch. Being from the country, I used to bumble and stumble and demur, but now I have learned that the only polite thing to say is, Well, the asparagus soup with wild mushrooms and truffle foam looks good.

First chance I get, I go for a run in Central Park, out around the reservoir. There are so many other runners, I feel like I'm doing a marathon. There is always someone faster than you, and always someone you can aim for and pick off. When I pass the little old lady in the powder pink workout suit I don't pump my fist or anything, but I find it helps the pace. When I am done, I walk awhile, through the tree-covered paths that wind up and down through the rocks and past statues. The heavy leaf cover mutes the noise of the city. For twenty minutes I sit on a bench with a view of the Manhattan skyline. A couple on a blanket in the grass makes me remember the shape of Anneliese in my arms. She is in

Colorado now, spending time with Amy and Amy's father, Dan. Dan has married, and he and his wife Marie have a little boy. Amy has a chance to play big sister. A few days after I go home from New York, I'm headed out there. From everything I hear, he is a good man, but I can't imagine it's going to be a day at the playground for either of us.

Back at the hotel, I check my e-mail and find a message from Anneliese. They have made it safely, all is well. While I'm online, I dial up the local paper so I can read Snook Ruud's obituary. For eighty years, Snook has been a fixture in my hometown. He was a shopkeeper, a butcher, a veteran, and an indefatigable raconteur, although if you were fool enough to trot out a term like that in his presence he would chortle and say, "More like a damn *liar!*" But he had that gift, where the stories just tumbled, and the old buildings rose again, and the village repopulated, and he could put you right on Main Street when it was all dust and Model-T tracks. Snook died just before I left for New York, and I am sorry to miss his funeral. Before departing, I wrote a brief note to the family and dropped it off with Bob the One-Eyed Beagle. He has agreed to hand-deliver it at the wake. Thirty years ago, Bob apprenticed under Snook. I imagine you can still see it in the way Bob slings a quarter of beef. I know you can still hear it in the way he tells a story. The old man was dear to him.

I navigate the subway, catch cabs, go to my meetings, and am taken to dinner. At Julian Alonzo's Brasserie 8½, I hold the menu with my knuckles because my cuticles still bear traces of grease from the last session in Mark's shop. I choose Cassoulet of Baby Clams and Roasted Loin of Rabbit from the prix fixe, in part because the rabbit comes with braised leeks, and the odds of my own leeks ever hitting the pan are dwindling in a forest of weeds as I sit here spooning up Peach and Blueberry Financier with Lemon Crème Fraiche.

I have one piece of New York business that is strictly personal. It is a pilgrimage of sorts. In an attempt to discover why photographs in the *International Light-Duty Series* sales brochure leave me wistful as a lovelorn schoolboy, I intend to hike over to the Whitney Museum of American Art for a chance to stand face-to-face with my favorite lonesome painting—Edward Hopper's *Seven A.M.*

I am no sort of art expert. I have only been to the Whitney once before, on a previous business visit that coincided with the Biennial. I enjoyed great swaths of that, although I was tempted to leave a note for several of the artists that said, "Great Start!" I would write it in crayon and add a smiley face so as not to seem rude. And I just do not have the patience for video installations, having yet to encounter one that conveys the absurdity of the human situation more effectively than a night spent channel surfing in a Motel 6 on the outskirts of Rapid City. But I like to look at everything.

I understand that *Seven A.M.* is not Hopper's most famous work. That would be *Nighthawks*, a stark existentialist portrait featuring a counterman and three patrons—two men and a woman—frozen in the timeless white light of an all-night diner. Perhaps you've seen the copycat version of *Nighthawks* featuring Marilyn Monroe, Humphrey Bogart, James Dean, and Elvis. The original *Nighthawks* resides in Chicago, and maybe (having never seen it in any form other than a reproduction) I am not giving it a fair shake, but as iconic and lonely-making as it is, I find it overpopulated. Superficially speaking, *Seven A.M.* depicts nothing more than an empty small-town storefront. But the first time I saw it—during that previous Whitney visit—my heart ached and I longed to crawl through the frame. I have done some reading since then, and know now that this was Hopper's ineffable thing, but I am happy to let him do it to me, and I have been waiting for a chance to stand before that painting one more time.

The *International Light-Duty Series* brochure is a twenty-page center-stapled full-color affair highlighting the L-110, L-120, and L-130 series. Published in 1950, it features gorgeous photos of nine truck models available in the series. The L-120 on the cover is identical to mine except for the color; in this case Apache Yellow. The palm fronds in the background are reflecting white light and appear slightly blurred, as if they are being pushed about by a sirocco. Perhaps the blurring is just a trick of poor printing, as some of the color separations are a hair off, but the taller shrubberies too appear as if they are bent by hot wind. Most of the photographs are taken at or around noon, with each truck standing atop its shadow like a place mat. The one exception—a Valencia Orange L-120 with a 127-inch wheelbase and stake body flatbed—is photographed on the gravel of a Midwestern farmyard. The photo is very closely cropped—you can see just a portion of one whitewashed brick outbuilding, a few square feet of the board-and-batten barn siding, half an open mow door, and no landscape at all—but based on the angle of the light, you can tell that the morning dew is still burning off. You can sense the coolness that remains in the air and the wetness of the grass that fringes the barn there where the white paint is blue in full shadow. In all the other photos, the meridian sun shows up only as a white glint along this or that convexity; beyond that the truck bodies reflect nothing more than color. On the orange flatbed, the horizon is visible in the door panel, and while the specifics are abstracted by the curve of the steel, the demarcation where dark meets light is sharp. You can imagine somewhere in the dark lower half of the door the cows dispersing to graze having given their morning's milk. The wind is yet to rise, and you can hear the farmer's boots on the driveway gravel as he turns away from the truck with a bag of feed over one shoulder. And when I look at that picture, everything I feel is, *I want to be there.*

○ ○ ○ ○

My walk to the Whitney takes me from mid-Manhattan, along the edge of Central Park, and into the East Side. The temperature today is in the high 80s, and the city is a swelter of activity. The fortresses overlooking the park are always a suggestive mystery, what with you being so close and yet so far, and I can't help but wonder what is going on behind each wall. If the cornflakes are stored in a sterling silver canister similar to the one discreetly shrouding the toilet paper and if mistakes are ever made.

Once inside the museum I go up the stairs and straight to *Seven A.M.,* place myself square before it, and just look. The sweet, sad pull is immediate. I dwell in it for a while, marveling at the trick that is being played on me, and not caring that it is a trick. I have been thinking about this painting for a long time now, and intend to soak it up.

The white squared-off building dominates the right-hand two-thirds of the canvas, cutting sharply against the forest that stands darkly to the background of the left-hand third. The plate-glass storefront is framed in tall white pilasters, and the outermost pilaster, the one fixing the corner nearest the trees, seems to bow out slightly. Perhaps it is a trick of the eye, but the impression is that the forest—bladed back from the building and viewed across a tan scrape—is exerting a vertical gravity. That the whole building is fighting the force of its arboreal roots. The forest is not at all inviting, and renders the building entirely alien, which, in the course of things, it is. Here in the age of the 'burb, we can infer that this forest, too, will be bladed back, its bulk and gloom nothing against the clackety-clack of the dozer, but implicit in its ominous primordial hue is the idea that man may impose flat planes and square corners on the earth, but entropy favors the organic form, and photosynthesis will win in the end.

The unsettling forest ballasts the painting with mortality, which only heightens the keen pining I feel for what it would be like to sit on a wooden chair alone in the room there on the second floor, the one with the wall of robin's egg blue. I grew up in buildings like this, and I know the feel of that lath and plaster wall, its coolness and solidity, the way your knuckle raps it solid, the way that sort of wall lends the room a sturdy echo, none of your brittle drywall skittishness. I can imagine looking through the window down to the street, the thin sheet of glass warping the view a bit the way old glass does. What I get is the sense of waiting, of stillness, and how it feels those times you rise with the sun and find yourself apparently in sole possession of the world.

Certainly some of what we've got going here is your garden-variety nostalgia. I read an article recently that said in round figures people have been gazing backward for ten thousand years. I am looking at *Seven A.M.* and longing for Harry Abrahamson's old country store as it still stood when I was a tot in New Auburn; I am looking at the picture in the truck brochure and remembering my own father hoisting feed bags. But I get the same helpless pang from the most disparate sources. The outro on Dwight Yoakam's version of "Good Time Charlie's Got the Blues"; a stone fence in Wales; the pale wallpaper in Oscar's apartment in reruns of *The Odd Couple;* the smell of tinder-dry pine needles warmed by the sun; the first notes of Liszt's *Liebestraum no. 3 in A-flat,* not that I would recognize the rest of it if I heard it in the dentist's office. Perhaps a better example is Marvin Gaye's "Inner City Blues (Make Me Wanna Holler)."

The atmospherics and Marvin's smooth angry voice make me wistful for a hot afternoon in 1971, which, given the context of the song and my rural white-boy roots, is on the face of it ridiculous. One velvety green spire of skunk cabbage alongside some airport runway and I hanker to trip lightly through the cowpies behind Daddy's red barn, where skunk cabbage stood in for the cactus of my cowboy books.

Not all the neural paths fire in such obvious sequence. The first time I saw those sun-blasted palms backdropping the yellow International, I thought immediately of chase scenes in *The Rockford Files*. I got hooked on reruns when I was in college. The theme music, the ongoing answering machine joke, Jim's put-upon wit, the way he ran like a stove-up ex-jock, I am fond of the whole package. But whenever he is in flight or in pursuit, my eye is drawn past the Pontiac Firebird into the background where California lies apparently lazy and hot beneath a sun whiter than the one we know here in Wisconsin, and beyond the set I see the new highways and the bare hillsides and I think of the subdivisions and teeming engines to come, and I become petulant over the fact that I can't wander in there. Never mind that the series was shot between 1974 and 1980 and we're hardly talking garden of the Hesperides. It's not about the preservation or the loss. It is that I have been cheated of that place in that moment. This is something beyond nostalgia and verging on *saudade*, a Portuguese word I first encountered in a Jim Harrison essay in which he spoke of obtuse sentimentality, childish melancholy, and a sense of life irretrievably lost.

I see Hopper's white clapboards and I know exactly how their paint will smell in the afternoon sun. Inside the blue room, I can imagine the seashell silence. If I understand Roland Barthes correctly, this is an example of *studium*—the attention we give a photograph because it contains elements that interest because of our accumulated personal experience and tastes. Shiny old trucks, gravel farmyards, a bag of feed. But Barthes also spoke of *punctum*, that part of the photograph that triggers something beyond simple recognition. Punctum is that element that arouses more ineffable emotions. In an essay bridging painting and photography, Peter Schjeldahl referred to punctum as a "quotient of inaccessible pastness," which seems right on the money. All the senses *evoke,* but visual images take you a step further. The visual image tells your heart, someone has actually *captured* this place, this space, and it is so close you can touch it, but access is blocked by the surface of the painting, a wall forbidding us to pass beyond our three dimensions. Barring the appearance of Mary Poppins, we are stuck here in the Whitney. But the real power of Hopper for me is that he has addressed the universe, saying, *Stop!* And it has.

○ ○ ○ ○

Before I depart New York, my editor takes me to lunch at the Monkey Bar. I become uncomfortable with the stares of the men in the four-figure suits, to say nothing of their companions: women apparently obtained on lease from some photo shoot intended to advertise a perfume the scent of which I can never quite place, although I suspect it is bottled in a slim decanter of frosted glass to which is affixed an embossed platinum label reading *Utterly Unattainable, Homeboy.* We were having the Roast Butternut Squash Soup and Seared Rare Tuna on a Bed of Steamed Spinach with Shaved Ginger and Plum Sauce. "Don't worry," my editor said to me sitting there in my Kmart socks and overworn T-shirt. "You've got it backwards. They're all trying to figure out just Who You Are that you got in here dressed like that. Act like you're Bruce Springsteen's favorite roadie." Sometimes these big-city people can be down home in ways that shame all your burnished small-town fables.

I was going to grab a cab for the airport, but several people have recommended that I reserve car service. A cab is cheaper, as I understand it, but if you get caught in traffic, or get a navigationally impaired cabbie, you may find yourself stuck with a prodigious bill. With a car service, you pay a set fee no matter how long it takes, and you can be confident that the driver knows the best route to the airport. So I make my reservation, requesting the simplest and cheapest vehicle, which in most cases is a plain black Lincoln Town Car.

I am waiting in the lobby when a driver appears and calls my name. Grabbing my shoulder bag and carry-on, I follow him out the doors. A gigantic white stretch limousine is double-parked and blocking the street.

I look ahead of it and behind it. No Town Car. The driver is popping the limo trunk, and now he is waving me over. I look at him with some confusion, but he beckons for my luggage, so I let him take it. As he draws open the passenger door and steps aside to usher me in, I shoot a look over my shoulder. A small crowd of tourists—all their luggage tagged by the same tour operator—is staring at me. As per usual, I am unshaven and wearing togs to match. I wanted to tell them, "Hey, I didn't ask for this . . ." But then I remember the Monkey Bar, and straighten up a little. Pay attention, folks, I am Possibly Famous.

I've never been in a limousine before, not even for prom. Very early in my writing career, I wrote an essay about the pretentiousness of limos and those who use them. And now here I am, pulling away from the proles in who-knows-how-many-feet of polish and chrome. How much am I paying for this? I wonder. From his seat one football field distant, the driver must have looked in the mirror and seen the look on my face. "Same price!" he says. "Same price!" I can't understand his accent perfectly, but I scoot up front to hear him better through the little window, and from what I can gather, the driver of my original car took ill and the company couldn't find a replacement in time. This limo was nearby and available, so they sent it.

Well, there you go, I think. This will make a great little story for the folks back home. I journey back across the football field and settle in. The driver's cell phone rings, and while he takes the call, I run my hand over the leather seat and stare out through the smoked windows at Manhattan in the rain.

Now the driver is speaking to me again. Telling me we must return to the hotel. Again, I'm a little fuzzy on the details, but he is saying something about someone else calling, about them renting the limo "for ten hours!" He is beaming. And why not. He takes a left, and then another left at Columbus Circle, and then we pull right back up to the hotel, where a standard black Lincoln now waits at the curb. Eight blocks and maybe three minutes after my grand departure, and of course the tourists are still there. *Downgrade!* they are thinking, as the Mystery Celebrity is hustled out of the limo to the waiting Lincoln. "Well, that was a quick trip," says the bellhop, raising one eyebrow.

○ ○ ○ ○

Back in New Auburn, I knock around like the bachelor I still technically am. My first night back, I go to the kitchen to make coffee before settling in to write awhile. I punch the button on the CD player and listen to Greg Brown sing "Steady Love" as I'm spooning out the beans. Anneliese bought the album after we went to the concert up in Bayfield. If you're going to be a bachelor, it's good to grind your own coffee and know a girl in Colorado loves you. I write until 2:30 A.M., when the fire department pager goes off and we drive through warm fog to a trailer house where an elderly woman gasps for air. We put the oxygen on, hold her hand, calm her, wait for the ambulance like we always do, and then it's back to the hall, where four of us including Bob the One-Eyed Beagle sit and tell Snook stories, one after the other, cuss words and all, until 4:30 A.M., when Bob has to go cut up beef.

I sleep a couple of hours, but am up at 8 A.M. doing laundry. When you have lived alone as long as I have, you develop some systems, and I am especially proud of my efforts to streamline wash day. It begins with the aforementioned Kmart socks. You may buy them at the discount store of your choice, but always make sure you get the ones that come packed six to a plastic bag, always buy the same style (I go crew-length), and always—this is critical—buy them in nothing but gray. By maintaining a strict sock monoculture, you eliminate the need for postwash sorting and can cram them all into the same plastic crate. In a rush? Grab any two socks from the crate and you are guaranteed a pair. The gray sock credo also eliminates the need for prewash sorting, as their standard grayness means I can safely wash them in the color load. Which is the only kind of load I ever have because I keep my closet clear of all whites, lights, and "bleeders" such as your reds and your maroons. A load is a load is a load. I call this my Unified Laundry Theory and you are welcome to it.

After hanging the last batch of wet clothes on the line strung between the back of my garage and the big maple by the alley, I peruse the raised beds. I have two big peony bushes out back, and by this time in July the leaves remain deep green but the flowers have pretty much had it. A few white petals cling, but the rest are just so much wilted brown crepe paper. This is why I don't care much for flowers. I like green things. Green is cool. Green is calming. Green hangs in there. V. S. Naipaul— whom I have read keeps an all-green garden—has been quoted as saying, "I feel if I wanted to see flowers, I could just take a bus ride and in front of every house there would be a series of shocking colours." Yessir. I keep the peonies because my mom had bushes that bloomed deep red and purple beside the steps of our white clapboard milk house and the scent takes me back, but that fond dalliance is quickly swamped by an overwhelming evocation of frumpy church ladies and their mysterious supertanker bosoms.

I pick a bundle of greens and, after washing them up in the house, make a salad with olive oil, canned tuna, diced black olives, fresh ground pepper, and a dash of white wine vinegar infused with chive blossoms picked from beneath the front porch window. Halfway through the salad I am thinking I might have gotten into the weeds some, because the back of my throat starts burning and eventually becomes downright uncomfortable. I keep waiting for myself to swell up or retch like a dog, but nothing really happens and half an hour later the discomfort is fading although my uvula still tickles. Boy, you just never know where the day is headed.

✿ ✿ ✿ ✿

Before leaving for Colorado, I get over to Mark's shop just once and don't get much done other than to stare at the truck and ponder the way the light plays off the ground sheet metal. There is still a lot of grinding to be done. Mark is excited because while I was gone he went looking for a part in a local junkyard and stumbled on an International with a cab that looked just like mine. The ornamental lettering said L-180 but the spec plate inside the cab said L-112. International and their sixty-six configurations of the L-Line—sometimes I think they themselves got confused. He said it was pretty shot but the fenders and grille looked decent, and he thought we could rob some parts off it. From what I can tell from my old books, the L-112 cab is the same as my L-120 and the L-180 looks similar, too, so I tell him it'll probably work. He says he's going to give it another look, take some measurements, and let me know.

○ ○ ○ ○

Colorado could have been trouble. In Colorado I put myself face-to-face with the past, not out of some sweet longing, but as a necessary step of knowing Anneliese and Amy. Anneliese lived in Colorado for two years. This is where she studied for her master's degree, this is where she met Amy's father, this is where Amy was born. Apart from my misgivings about meeting Amy's father, I knew our visits with Anneliese's college friends would be laced with reminders that I was being written into a preexisting story line. In this respect, old friends always have the upper hand on new lovers. Three months of pillow talk do not supplant the ratty sweatpants of history. This is an issue not of speed but mileage. You wonder if you'll hit it off, you wonder how you're stacking up, you wonder if this chair in the coffee shop has been occupied by a previous object of affection. It is my pet theory that men are much more childish about this than women, or certainly I am. In my past roles as lover I have behaved in ways that left me accused (often rightly, sometimes wrongly, but always at cost) of blockheadedness, contradiction, and worse, but with each new relationship it always seemed to be I—not the woman— who went through a stage of obsessing over the past. Running little films of scenes I had neither right nor reason to conjure, and yet like the pro-verbial sore tooth revisiting them again and again. As if I could jump in and intervene. Sometimes it was all I could do not to remonstrate my partner for doing things I had done fourfold. Grievance is a sullen little boat, blown in the creepy breeze of ridiculous sighs.

We root around in the past because the future is unavailable. It is harmless enough, I think, at this point in my life, to stare at old truck advertisements and wish to be somewhere I cannot. To triangulate be-tween the Hopper painting, the International ad, and the current state of my heart. We sort the past in an attempt to sort the present and antici-pate the future. I am paying to put new paint on an old truck in part so that I may use it in the present, but I am also trying on some level to pick the lock on Hopper's quiet blue room. I buy copies of *Freezer Fancies* or *Prelude to Home Freezing* so I can gaze at Irma Harding for the fun of imagining all she was meant to project, but I am also catching a little frisson off the pages of a cookbook that speaks to me from the kitchen of some fifty-odd years ago. Other people will pay $186.08 for a box of Irma Harding tin foil, $179.05 for an Irma Harding timer, and $87.55 for a flyer advertising Irma Harding freezer packaging supplies. I recently purchased an advertisement torn from a 1951 edition of the *Saturday Evening Post*. It features a picture of an L-model International pickup and text declaring, "What you pay isn't half as important as what you get." The hope is that by inhabiting moments that are unavailable—because they are in the past or never existed at all—you will be arming yourself to recognize the real thing in real time. That you might recognize the moments you long for when they are happening.

The downside, of course, is that you can auger in. I say this as one who tends to wallow. In preparing for the Colorado trip, Anneliese and I have been quite naturally led to revisit our separate histories, comparing tote boards of regret. Anneliese is matter-of-fact in these matters, whereas I adopt the demeanor of a consumptive poet, heaving my chest weakly, construing all manner of mournful torment in what cannot be undone. When she has had enough, Anneliese speaks to me at a level I can understand, which is roughly Dr. Seuss: *We are what we are because of what was.* That is to say, *What you pay isn't half as important as what you get.* She is, of course, clearly stating the obvious, which by now I assume she realizes will be a regular requirement of hanging out with me.

Sometimes it is the future that calls out the past. In talking about what might be going on in Amy's little heart and head during the visit, I told Anneliese how I have come to love the little girl and tears sprang to her eyes with an immediacy that left me spooked. I thought about my young friend Adam waiting for me to take him fishing again, and the litany of my disappointed lovers, and I got sick with myself at the idea that I might be the alcoholic who says he has put the bottle down for good.

It was a relief then, that Colorado went just fine. I was able to assign faces to stories. I was able to meet several people who were at Anneliese's side in the difficult months surrounding Amy's birth. Amy's father, Dan, turns out to be a humorous and articulate fellow and although there was potential for an episode of reality television, the six of us spent the bulk of the week under the same roof and we never once pelted each other with empty beer cans. On the third morning or so, I found myself alone at the breakfast table with Dan's wife, Marie, and we had our own little heart-to-heart based on the parallel elements of our respective roles. I remember leaving the table thinking, We can do this. It took time and hard work for Anneliese and Dan to reach this point—there were gaps, and both can claim their scars—but the result is that for six days, we all gathered at the same supper table, drawn together by a child who currently responds only when addressed as the great racing horse Seabiscuit. For the purpose of balancing all the happy talk, I should say that during a visit to the Denver Firefighters Museum, Amy went off on a fit the length and breadth of which deceived several eager museum patrons into thinking they had arrived late to a live reenactment of a historical five-alarm clanger.

○ ○ ○ ○

I return from Colorado to find all the tomatoes in the backyard dead or shrunken. I stand there staring at them and think I should write a gardening book and call it, *I Know Why the Caged Tomato Sags*. The first year I planted tomatoes back here they went like gangbusters. Every year after, it seems half of them die. They roar up and get to a point where they look full and green, and then they develop this habit where by noon they wilt. If I jab holes in the ground and pour water to them, they come back, but it's short-lived. Someone told me it's a root ball problem, and someone else told me it's the black walnuts on the property line. I know black walnuts will kill tomatoes. But I had that one good year, and so I keep planting some back here. It seems like the cherry tomatoes survive better than the larger breeds.

The good news is, the tomatoes I planted in the new bed between the house and garage are steroidal in their abundance. My mom stopped by while I was gone and picked the ripe ones and left them on my porch. The plants are so huge they have tipped over and upended the wire cages. Must be the pig manure. I drive in stakes and tether the tomatoes to the stakes to hold them upright. With my usual penchant for overkill, I planted cucumbers and beets in the new bed as well, and the cucumber vines are swarming the beets. I just don't get the concept.

Out back, where the big maples cast shade across the yard for most of mid-afternoon, not much is happening, although I have noticed the slow pace suits many of the herbs and all the lettuce, which grows at a pace you can keep up with instead of bolting. And the peas have done well, having reached the top of the trellis, where they have turned and have begun fountaining back to the earth. But the cucumbers back here are lousy, all spindly and tentative. Only two of my basil plants are big enough to harvest, and they look a little pale, but I snap off the best leaves, enough to make a batch of bruschetta for lunch. I make my way around my meager little plot, gathering cilantro, parsley, and a few green onions. Then it's to the sink, wash and chop, the leaves looking greener against the wood of the cutting board. In addition to the ingredients I have picked, I dice several cloves of garlic. Everything goes in a ceramic bowl, then I drizzle in virgin olive oil, red wine, and fresh-squeezed lime or lemon juice. I mix it well, and cover to sit out at room temperature until lunchtime. I add fresh ground pepper and grated parmesan late. When it's time to eat, I boil water and make a batch of angel hair, strain it, and then stir in the bruschetta. I don't know if this is proper or not, eating it warm, but it seems to me it gives everything a fuller flavor.

I eat in my favorite spot, the big green chair in the living room beside the bookcase with a view through the screen to Main Street. I can't imagine a finer moment than to be here in this old chair with this fresh alive food in my lap, all the greenness and the garlic and the sounds of the day easing through the screen on the back of a breeze. The bruschetta recipe comes from an e-mail printed and pinned to my recipe board. It's from the poet Bruce Taylor, an above-average hedonist who once stood by an open window in a bar on a spring afternoon and said, "Sometimes the best thing to do with a beautiful day like this is to spend some if it sitting in here looking out." There is something about listening to a day through a screen that infuses the moment, as if the steel mesh slows the day down, lets us bathe in it a bit more. A screen seems to filter the harshness from the outside noises and they reach your ear softened. It will be best if the sound is coming to you over a varnished wooden floor decorated with a strip of sunlight; the flat surface, however artificially imposed, is reassuring in the face of entropy and has the added advantage of being made from trees and blessed by light. It is exquisite to sit here in this perfect moment, eating food that I—a black-thumb gardener—have coaxed from seed to fork. I am humbled that in the face of all chaos, I should have this plain, priceless moment.

And then the nap. Set the bowl on the floor, tip the head back, take the glorious option of not fighting the heaviness in each eyelid. Maybe you shift your shoulders a little to get just right, and then there you are, sleeping sitting up in the middle of the afternoon of a perfect day. If you ride the wave right, catch it on the downslope, snag that catnap where you dip into unconsciousness and then rise smoothly back to wakefulness after only a few minutes, yet having shut down long enough to defragment the mind, O, then that is a glorious thing not to be replicated with any long snore. You come awake with freshness and clarity and the strip of sunlight has shifted, and you are living punctum in the present, *saudade* before it is sad.

CHAPTER 9

AUGUST

To the best of my knowledge, my brother John arrived at the age of thirty-five having never been on a date. Then one morning last year the phone rang and it was my mother.

"We think your brother has a girlfriend."

Pause.

"How did *that* happen?"

My brother lives in a tiny log cabin surrounded by jackpines. His only vehicle is a dump truck. Honestly. If you need some rocks hauled, he will haul them with his dump truck. If you need a few yards of black dirt, he will deliver them in his dump truck. If you invite him to dinner, he will arrive in the dump truck. So when I heard he had met a woman, I was flabbergasted. It was like waking up one morning and finding a fifth face on Mount Rushmore.

"How?" I asked again.

"Well, as it turns out," said my mother, "she drives her own dump truck."

○ ○ ○ ○

Come August, you feel it all slipping away. The garden weeds are seeding out. The tomatoes ripen faster than you can figure what to do with them. You force boxes of surplus zucchini on complete strangers. You realize the leeks simply are not going to turn the corner and will remain the diameter of Tinkertoy sticks. A handful of the hottest days of the year are yet to come, but some afternoons the sunlight is dilute and fails to heat the air, which in turn hits your nostrils with a remindful zing. This week I was on the front porch steps lacing up my running shoes in preparation for yet another four-mile slog when the giant maple across the street produced an eruption of blackbirds flying outward and apart. On some invisible cue the birds pitched, cohered, and streamed directly overhead with an ominous feathery hiss. It sounded like the air was colder up there. These were redwings, and they chuckled in a way that reminded me of frogs in spring. Twin sounds, bookending summer.

When I say my brother John lives in a tiny log cabin, I do not intend "tiny" as a euphemism. If you walk through the front door and stop, immediately to your left you will find a small wooden table, which he made himself. If you then progress from the table along the walls in a clockwise manner, you will encounter a stove, a sink, a refrigerator, a wood stove, a washer, a dryer, a hot water heater, a chest of drawers, a bathtub, and then you're back where you came in. A long-armed man with a pair of pasta tongs could stand in the center of the room and pretty much run 'em all. A hinged ladder beside the tub leads to a mattress stuffed in a cubby hole wedged between the beams and purlins. Some might call it a loft.

John and my brother Jed cut the logs and built the cabin themselves, constructing it in the yard beside the machine shop on Jed's farm. It was a project for Sunday afternoons. The original plan was to get it built and sell it to a tourist. Then John bought a patch of land on the far side of the big swamp off the back of Jed's farm and decided to keep the cabin for himself. Problem was, his homestead was two miles west of Jed's shop as the crow flies, with several hundred acres of impassable tamarack swamp in between. The journey by county road ran a good three miles. Small as the cabin is, it would hang well over the centerline. There are rules about these things. To transport it legally they would have to rent a truck, hang signs, and navigate a government permitting process culminating in the writing of checks. On the other hand, John says if you get caught hauling something like this behind a tractor, you can get off the hook by claiming you are just a dumb farmer. Jed fetched the Massey Ferguson.

Then they raided the iron rack for some goodly lengths of channel iron. These they cut up and welded back together in a T-shape of a width allowing them to hitch two hay wagons in parallel formation behind Jed's Massey. Employing a conglomeration of jacks, they raised the cabin in slow sequence, inserting blocks until the wagons could be slipped beneath. Once both wagons were positioned, they gently lowered the cabin. The wagon beds creaked but held. Then they went home to their beds. Very early the next morning, Jed took the whole contrivance on a practice lap in the field behind the barn, just to make sure she would track and turn. Then John pulled in behind and chained the front end of his pickup to the wagon frames. Hay wagons are brakeless, and were Jed to hit the tractor brakes, the tandem wagon hitches would likely jackknife and dump the whole works in his back pocket. It would be up to John and his truck brakes to hold the cabin back. He would do this mainly by intuition, since the view from his windshield was blocked by a wall of logs.

There are mobile homes, and then there are mobile homes. What you had here was a double-wide float in a hillbilly parade. You could have shingled the Department of Motor Vehicles with the citations required to summarize the moving violations committed the moment the first wagon wheel touched blacktop. When my brothers retell the story nowadays, they grin and admit they were nervous. John says the second they pulled safely off the road and onto his land, he said, "Hey—let's not do that again!" But they made it. Due to some wedging, they had to dismantle the wagons to get them out from under the cabin, but eventually they got the thing planted level. John had a home.

The cabin was wired with electricity and plumbed for water. But the bathroom was located out back, in a separate building. Specifically, a small wooden hut with a crescent moon cutout in the door. There are times, John says—particularly on January mornings—when it is difficult to muster the courage required to throw off the blankets and make the trot.

In addition to his dump truck, my brother owns a skid-steer, a track hoe, and an equipment trailer to haul them. He's one of those guys, if you need to put in a driveway or dig a basement, you give him a call. Eventually he'll show up with his equipment and his dog, Leroy, and do the job. I say eventually because he stays plenty busy, and if you call more than twice in one week you reveal yourself as a city-bred newcomer and should repair to your collapsible canvas chair and reflect on the fact that you moved up here to relax. You may wish to emulate Leroy, who will mask his enthusiasm for extended excavation by snoozing in the cab.

As I have come to understand it, John and Barbara—the woman with the dump truck—had been seeing each other for years. When I say "seeing each other" I don't mean clandestinely dating, I mean casting and averting gazes and pondering possibly maybe making moves. Eventually (that is to say, "about five years down the road") Barbara quite correctly deduced she would have to make the first move. One day after delivering a load of gravel to a site John was working, she climbed down from her truck cab and approached him. He was pushing dirt with his skid-steer, and she had to wait for him to kill the engine and tug his earplugs free. Then Barbara said, "If I asked you to go to dinner with me, would you consider it?"

Later, Barbara said John wrinkled his nose and appeared stricken with gastric reflux.

"Ahmm . . . maybe."

Yes. *Maybe*. Jiminy. When Barbara climbed back in her truck she figured she had blown it. The next day she left for a ten-day vacation, thinking she could never deliver gravel to this man again. Within a month, he ordered another load. She dumped the gravel, and as she was pulling out he flagged her down.

"You ever folk dance?"

"Erm."

"Next Tuesday, I'm going folk dancing. You wanna go?"

"Well . . . sure."

She said she drove straight home and got on the Internet. "I had to see what these folk dancers wore."

○ ◎ ○ ◎

For the most part my brothers and I do each other the favor of recus-
ing ourselves from each other's personal lives. In the face of tragedy, or
when specifically invited, we are there foursquare. But by and large, we
limit our demonstrations of affection to flashing goofball salutes or peace
signs when we meet rounding a curve on Old Highway 53. We got the
peace sign thing from Dad. In his farmer overalls and Lil' Abner work
boots he is the least tye-died of men, but for as long as I can remember
he has greeted friends and strangers alike with the peace sign. Rather
than flipping it up lightly beside one ear, he pushes it toward you in a
thick-fingered vee that resembles a sprung bundle of sausages. The complete absence of panache suggests that he really means it. Either that, or
he's channeling Winston Churchill.

If we meet on the road and time allows, we'll stop for a centerline
chat. Add this to the list of rural traditions on the wane. What a delicious
refutation of hustle to align your driver's side windows, kill the engines,
and shoot the breeze while the flies buzz. You talk about where you're
headed, where you're coming from, how the corn's looking, or the price
of hogs. You keep one eye out for traffic. If a car approaches and you can
wrap it up, it's crank the starter, roll off, and toss a *see-ya-later!* over one
shoulder. If you're in the middle of a good part, you pull ahead enough to
let the traffic pass, then back up, realign, and pick up where you left off.
Usually it's just chitchat and catch-up, but sometimes you get nuggets. It
was through pickup windows at the intersection of Carlson Corners that
I received the happy news of Jed's engagement to his second wife.

The land rush is on in these parts, and not all the new folks are pursuing a ruralist vibe. They roar up here and forget to quit hurrying. This
past summer I was mowing the lawn when my buddy Snake passed by.
I flagged him down, killed the mower, and wandered out to talk. Snake
and I were pals from kindergarten to graduation. These days we see each
other maybe once a year. I leaned against his door there in the middle of
Main Street, and we visited for a good while. Every now and then a car
would swing around us, but you can run four or five abreast down Main
Street, so it was no big thing. Then this woman pulled up on Snake's

bumper and honked. I looked at her, looked at the space around us, and
then leaned back in the window. She honked again. We just talked and
ignored her. Shortly she gave the steering wheel a violent twist, stomped
the accelerator, and whipped out around us. As she zoomed past, she
gave us the finger. We gave her the gaze. The implication being, Ma'am,
this is how you gossip in the absence of a garden fence. We are luxuriating in the tapering moments of a quieter time, and furthermore, honking
crabs the soul.

During the early days of his relationship with Barbara, John and I chat-
ted on the road several times. He'd either clamber down from his truck
to my level, or I'd jump up on his running board and hang by the grab
bar. As hopeful as I was for their happiness, I never asked him how it
was going. I knew they were seeing each other regularly, and I met her
a couple of times at family functions. She was pleasant, smart, and attractive. When she wasn't hauling gravel, she ran her own tax account-
ing firm. And she had a beautiful truck. A big red Mack. Finer than any
piece of trash John has ever run. But as to the state of the relationship,
I didn't inquire.

Several months into the deal, I had to borrow a tool from John. I
drove out to where he lives and walked up the trail through the jackpines
to his cabin. He was setting up forms to pour footings for a small addition. "Whad'ya doin'?" I asked.

"Ahh . . . puttin' in a bathroom."

Oh-ho! I thought.

But I didn't ask.

$$\circ \; \circ \; \circ \; \circ$$

It's funny to think of him folk dancing, because my brother has a reserved stiffness to his comportment. (Then again, much of folk dancing
is quite nicely prescribed, and furthermore, he was lured into this particular vein of decadence by my mother, a churchly woman whose epithet
of choice is *fiddlesticks!*) Sometimes when he laughs, he squints up his
eyes and tucks his chin toward his collar in a manner that makes it look as
if he is embarrassed for getting so carried away, and if you can just bear

with him for a second here, he will straighten up and get it together. He and I come from a long line of Scandinavian stoics. In most social settings we are, if not shy, determinedly reticent. Our guiding precept is, "I don't want to talk about my feelings, *and you can't make me.*" My tears have loosened a bit with age, and I can be grumpy, and scowly, and—as my mother used to say—*a little snippy,* but in general I am pathologically self-contained. Writing-wise, I share things on the page that would mortify me if they came up in casual conversation, but these seizures of self-disclosure are triggered by the imminence of tongue-loosening deadlines and vertiginous health insurance premiums and should therefore not be confused with me at the post office, where I tend to study my boots and mumble.

Would that it ended there. In the world of the certifiable stoic, the repression of emotion is just the more obvious half of the battle. The rest of your time is consumed with masking even the appearance of the existence of desire. Anyone can hold back a tear or dodge a hug—it takes a real hardcore Norwegian bachelor to pretend you don't want a cookie. If I were commissioned to design the official crest for the descendants of emotionally muzzled Vikings everywhere, I would begin by looking up the Latin phrase for "No thanks, I'm fine."

This outgrowth of the neurosis turns the simplest trip to the grocery store into a pulsating gauntlet of dread. Shopping for staples seems benign enough, but when you present your basket at that counter, you are revealing something deeply personal about yourself. You are approaching a stranger and saying—*in public*—"this is what I desire." And not only that, but, "this is what I desire *to put inside me.*" If you are buying a battery cable or a snow shovel at Farm & Fleet, there is no shame. These are exogenous needs. Gotta start the car, gotta clear the sidewalk. But with food, there are distressing elements of psychosexuality in play—*Appetites! Hungering! Orality! Gimme Twinkies!*—coupled with the implication that if you ingest you must surely excrete, and this is not a place the stoic wants to, um, *go.*

Furthermore, one risks public exposure by checkout clerks who take it upon themselves to deliver unsolicited color commentary on the contents of your cart. I recently skulked through the IGA to snag a box of broasted chicken when I should have been cooking at home, and at the register the lady said, "MMMM, that smells good!" and right she was, but I immediately felt as if I was standing there in my underwear. The same thing happened when I tried to unobtrusively purchase a transparent plastic clamshell of chicken tenders during a late-night road trip. "Oooo," said the young lady running the register, "I like those reheated."

Frankly, I didn't know what to think, but I did get vaguely sweaty.

Sometimes it's cumulative, like water torture. Last fall I got the urge for boiled dinner, which I associate fondly with crisp air and my mother's cast-iron wood stove. As I riffled through my checkbook, the play-by-play commenced.

"Celery! Yum!" My abdominals tighten.

"Carrots!" My forehead is beginning to prickle.

"Onions!" *You recognize them, then.*

"Rutabagas! Oh, I like rutabagas! And cabbage!" I now have visible beads of scalp sweat.

By the time she swipes the smoked ham hock across the scanner and holds it up like she's Liberty Enlightening the World Regarding Cured Meats, my heart is going like hummingbird wings, and my skyrocketing blood pressure is causing an incremental protrusion of my eyeballs, an effect similar to those pop-up timers that tell you the turkey is done. At which point, in a voice that can be heard clear back to the deli department, the checkout lady announces, "LOOKS LIKE SOMEBODY'S MAKING BOILED DINNER!"

I wouldn't turn any redder if she pantsed me.

It's far worse when I'm buying something dietetically naughty. As a fellow who has been known to run nine miles up the road to Kwik Trip at 3 A.M. to score a twin pack of Little Debbie Oatmeal Creme Pies and an extra large Royal Kona Blend, I am deeply grateful to the checkout person who understands that what we have here is the equivalent of a drug buy, and both parties shall honor an implicit commitment to dispassionate efficiency.

Omertà on the Zebra Cakes, as it were.

Fulfilling all suspicions, a couple of months after the construction of the bathroom commenced, John and Barbara announced they were engaged to be married. It turns out Barbara had agreed to join John in his rustic shoe box, but she predicated the move on one very specific stipulation: she would not sign the wedding license until John installed an indoor toilet. The wedding is two weeks off. Everything (and by "everything" I mean to include the license, the cake, and the sewer line) is in place, but the toilet has yet to be connected.

John's first public announcement of pending nuptials was made during the "new business" portion of the monthly meeting of the New Auburn volunteer fire department. I was sitting in the back row of folding chairs. There was a flurry of people straightening in their chairs. Some of the guys shot each other elbows, and *Oh-HO*'d and *oooh*'d in his direction and made him turn red. Another Perry boy off the list. And then, just as quickly they all turned and looked straight at me there in the back row. All these years, and I am the oldest brother and last standing bachelor. They might as well have thrown me to the floor and tattooed my forehead with the words *NEXT PROJECT*.

○ ○ ○ ○

The real wild card with any stoic is rage. By and large, I am the least wrathful of men. But there is within me a vexatious little ball of propane, spritzed with paint thinner, lashed to the tip of a sulfur match, and hidden beneath a pile of oily rags just to the left of my spleen. When specific triggers are tripped, I fly apart like rivets off a tin flywheel. Thankfully, this rarely happens in public. Excepting the tantrums of childhood, and an incident in Wyoming in which I was caught barking at a rototiller, I have remained a closeted rager. In public, I prefer to keep a cork in it. But absent witnesses, I will let fly like a goose exiting a turboprop.

Any number of things will do it. An open cupboard door to the forehead; dropped Internet connections; bookshelf kits short one screw. Some of my most vicious unhingements have erupted during solitary forest strolls. It seems counterintuitive, what with the restorative aspects of nature and whatnot, but try it sometime when it's zero degrees, your cheeks are stiff as lard, and you get snapped in the face by a sapling switch. A clunk to the head is bad enough, but the impudent smack of a sugar maple sprig is akin to the flick of a doeskin glove from some ruffle-throated dandy. I go from Thoreau to Mr. Hyde in a nanosecond, provoked by the stroke of a branch no bigger than a flyrod tippet. I'm glad you didn't see me last fall, one watery eye clamped shut, in full-out wind-milling flail, varnishing the remains of a birch sprig with spittle and curses. Far above, from the safety of a sturdy oak, a squirrel chattered and wheezed, quite rightly perturbed at the presence of a sinner in the forest.

Back in the civilized world, my own stupidity is a regular flashpoint. I once responded to missing an important meeting by hurling the top half of a papa-san chair across the room like a gigantic Frisbee. I can report that it splintered against the wall with a sharpish ker-*rack!* and I felt immediately improved although a little down in the mouth over the destruction. Later, having smacked my elbow on the file cabinet for the second time in five seconds, I grabbed the nearest unattached object, rared back, and flung it with all the violence I could summon. The object, as it happens, was a 16-count Kleenex Pocket Pack. I cannot recommend the results.

And the language! Well, never mind the language. Suffice to say it would flabbergast those who love me, as I am characteristically easy-going, and prone to utterances no more scabrous than *jeepers* or *dangit*. But what I'm thinking is, should I survive to senescence, I'm bound to be one of those sweet old folks who winds up slinging stewed prunes and cussing like Ozzy Osbourne doing a Mamet play.

◇ ◇ ◇ ◇

At 10 A.M. on the morning of August 23, we gather on the shores of Round Lake for the wedding. John finished installing the toilet at 5 P.M. last night. He called Barbara, essentially to tell her the toilet was in and the wedding was on. She wasn't home. He got her answering machine. After the beep, he said, "Yah. I've got a message for you." Then he held the receiver down by the porcelain bowl and hit the flush.

The service is brief and sweet. We gather up around them, standing quietly around the boat landing, where you can hear the water lapping.

I can see John's hands have a tremor in them, something to see in a man who is the best rifle shot among us, and who even when we were kids could squeeze the basement scale and spin it past anything the rest of us could manage, including Dad. It's fun to tell bachelor stories on him, and cast him as this rough-hewn throwback, and it's all true, but when I see him there trembling a little, I am seeing the brother who learned sign language and used to sit in the classroom beside a local grade-school girl and translate her lessons, which must have looked like Grizzly Adams meets Little House on the Prairie. I see the kid who read all the Foxfire books and taught himself to tan hides and make wooden door latches, and who kept cranking the draw weight on his Browning compound bow up tighter and tighter until one day he pulled it back and it exploded. And while I see a man I sometimes envy for his ability to build things and fix things and run things, I also know that last year during the community choir Christmas concert he stepped forward for a solo that culminated in a note so high and pure it put my heart in the rafters.

Barbara looks strong and maybe a little nervous, too, but when they speak their vows in turn you can hear them considering every word. There is no preacher, just a judge, and when it comes time to sign the papers, the judge is at a loss in the open air until John turns and offers his back, and as the judge scribbles across his shoulders, you hear a scatter of laughter, and that's a nice way to end as we walk back up to the picnic tables where all the kin and old neighbors have gathered, and the Nesco roasters are lined up and in the air you catch the scent of charcoal chicken, barbecued and served up hot by the New Auburn Area Fire Department. Sometimes life is so simply good.

After all the usual photographs were taken, John and Jed and I stood shoulder to shoulder for a portrait with our three left hands extended toward the camera, palms in and fingers spread to show three hands and two rings.

○ ○ ○ ○

It takes until the end of the month, but I finally make it over to work on the International. My arrival coincides with that of a flatbed truck delivering the International Mark discovered in the junkyard last month. He took his measurements and I looked in my books, and we decided it will work. It appears the truck was originally Chesapeake Gray, but the fenders are furred with lichen—at first glimpse it could be the same strain that grew on my truck. It was once fitted with a wooden flatbed, but all the planks are rotted or removed, leaving only the square channel-iron framework.

If the license plate renewal stamps are an accurate indication, the truck was taken off the road in 1975. The plates are old-school Wisconsin style, lemon yellow with black numerals. Inside the cab, the bench seat is still in place, but mice have devoured a big chunk of the upholstery, leaving behind seat and foam croutons. Over by the passenger door I can see an old plastic take-out basket full of fired center-fire rifle cartridges. I take a guess: "30–06?" Mark digs one out and checks the imprint on the rim. "Yep."

The cab is full of these distracting treasures. A notebook with notes toward an indeterminate engineering project; a 1974 edition driver's education manual; three maps, one of Idaho, one of Montana, and one combining Nebraska and both Dakotas; a canvas carpenter's apron with nail pouch; the toe binding from a downhill ski; various matchbooks; a receipt for nine and a half yards of concrete; a savings passbook indicating that in 1975 a man named Lester had $9,088.66 in a local bank. These disparate threads of someone's story are seductive. One minute you are checking out the state of the accelerator pedal, the next you are trying to fill in the blanks. The simplest object is immeasurable. Every footnote has a footnote. I am thinking again now of Borges and his Garden of Forking Paths, his infinite Library at Babel. I am thinking sometimes a little Ritalin might not hurt a guy.

Mark is under the hood. He says the fuel pump is gone, but it looks like the carburetor is the same model, so we may be able to rob parts from it sometime if need be. Meanwhile, while scavenging the detritus in the cab, I notice that the floor is covered by a rubber mat. At first I figure it's something a previous owner slapped in there, but then I clear away a water-stained parts box and there between the shifter and the firewall is the classic Raymond Loewy IH logo, embossed right in the rubber. Without meaning to I say, *"Oh!"* probably just like Kathleen did when she found her engagement ring planted in the toolbox.

Solid and square and formed so simply of the two sturdy block letters, the IH logo was first used on tractors in 1945. By the time my
beloved L-Line debuted, it was in use throughout the company, ap-
pearing on over seventeen hundred items—everything from boxes of
Irma Harding–approved freezer paper to the distinctive pylons jutting
from the Loewy-designed International Harvester Servicenter buildings
dotting America. Known as the man who streamlined everything "from
lipstick to locomotives," Loewy came to America from France after fighting for his country in World War I, during which he demonstrated his
commitment to esthetics by hand-tailoring his uniform and draping his
trench with fabric. In a career that extended into the 1980s, Loewy had
a hand in a mind-boggling number of projects ranging from the iconic
Lucky Strike cigarette pack (bet $50,000 by the president of American
Tobacco that the packaging couldn't be improved upon, Loewy collected
in short order by simply placing the distinctive bull's-eye image on both
sides of the box, so no matter how the carton was tossed, the logo landed
up) the Greyhound Lines logo (he replaced the original "fat mongrel"
with a lean silhouette approved by the American Kennel Club), and the
Exxon logo.

By his own account, Loewy designed the International logo on the
back of a dining car menu on the train taking him to New York after his
meeting with International in Chicago. Declaring the previous International logo (the letters IHC stacked on top of each other and enclosed
in a circle) "frail and amateurish," Loewy centered a lowercase *i* over
an uppercase *H*, drawing for his inspiration the image of a farmer on
a tractor: the dot his head, the two legs of the black *H* representing
twin drive wheels. "Before we passed through Fort Wayne," he wrote,
"International Harvester had a new logo." For all that has been written about Loewy's sleek contributions to redefining industrial form (he
once designed an aerodynamic pencil sharpener), he is quoted as saying
that he preferred simplicity over streamlining. Perhaps this was a tad
disingenuous, like Edward Hopper saying his critics had overdone "the
loneliness thing," but it sure enough worked for the International logo,
blocky as it is. There are some things that just sit right with your eye, and

that logo is one of them. That logo doesn't say streamline to me. It says boots on the ground.

I call Mark around and we carefully peel the mat from the floor, then lay it out on the concrete. It's a little crusty and dusty, but appears to be in perfect shape. I hose it down and then scrub it with a push broom. It looks factory new. Gorgeous. We stand there looking down at it, grinning. Every time I climb in my truck now, I'll be able to look down and see my favorite monogram in the world, right at my feet. Courtesy of a man who designed forks for the Concorde and the toilets on Skylab, and who combated the distasteful odors in his deep-sea diving helmet by adding Chanel No. 5 to the air-pumping mechanism.

The floor mat is a delight and a coup, but our primary purpose today is to rob the fenders and grille from this truck so they may be grafted to mine. You keep a junker like the L-180 in the weeds out back and "part it out" as necessary, cannibalizing it whenever you're short a door handle or a mirror or a distributor cap. Kneeling at the front bumper, we give the grille a once-over. The L-Line grilles reflected a major change in design, and remain one of the easiest ways to distinguish the trucks from their ancestors and offspring. The horizontal grille strips of previous models were replaced with vertically oriented slots overlain with two horizontal bars. In collectors' circles, this is known as the mustache grille.

The grille on the L-180 isn't perfect—there's a fair amount of rust around the headlights, which someone has repaired with tin and rivets—but it's in better shape than the one on my truck. Right off the bat I notice a slotted hole lowdown and dead center and recognize it as the insertion point for a crank starter. In the absence of battery power, you just stuck the crank in there and turned the engine over until it caught—or backfired and broke your elbow. The old-timers will tell you these stories. One of the horizontal bars is badly warped, and one is straight. The good news? It's the same story on my truck, so we'll use one straight bar from each.

Before we can pull the grille, we have to pull a homemade brush guard. It's fairly roughly done, cobbled together out of welded quarter-inch steel and bolted to the bumper. We're getting ready to loosen the bolts when I notice a gap missing from the center of the guard. It's a crude cut, punched through with a blowtorch and left rough. The serrated path of the flame is still visible. But what makes us smile is that it's centered right over the starter crank aperture. We stand there a minute and enjoy the idea of it. The guy working all afternoon, welding the whole works together, drilling the bolt holes, grunting it into place, snugging all the bolts down, and then—in that signature moment of maleness—taking two steps back to pause and admire what he has wrought. It's the same when we finish shoveling the driveway, or stacking wood, or folding a dish towel. But in this case he basked there for a minute, and then it slowly dawned on him that he had blocked off the aperture for the starter crank. You can just see him in your mind's eye, shaking his head and going for the torch.

On the other hand, maybe he didn't discover his mistake until the first time he tried to start the truck. Again, you can see him standing there as his jaw loosens and drops into the *aaacch!* position, then he throws down the crank and spins on his heel.

Either version, we get a kick out of it.

By the time we get everything detached and laid out on the concrete apron, the mercury vapor light over the shop has come on and the crickets are commencing. By chance, I read a Stanley Kunitz poem over lunch in which he had a line about crickets *trilling/underfoot* and it's nice to recall it now. Mark and I shoot the breeze easy for a while, and then I drive on home.

Later that night the cops get into a car chase that ends in our county when the pursued driver crashes his Corvette into a tree. The tree is barely scuffed, but the nose of that car is in pieces all around our feet, and I am thinking, *Son, what you need here is one of them International mustache grilles.* What is to become of a country that trades steel for fiberglass? In a bit of delicious irony, the man credited for recommending Corvette go the fiberglass route was Raymond Loewy.

❍ ❍ ❍ ❍

I get back over to Mark's place early the next morning. I prop the grille up against the shop and start drilling out the rivets holding the tin patches in place. The ridge formed by the edges of the tin overlay were smoothed over with body filler, long since cracked. When the last rivet is drilled and I pull the tin free, I find that once the filler cracked, the potential space between the tin and original body became a moisture trap, and the rust is more advanced behind the patch than it would have been if left to open air. Even so, the deterioration is nowhere near advanced as that on my truck.

The headlights are still in their sockets, so I loosen the screws and pop the retaining ring, and when the sealed beam comes out in my hands, the paint behind it is factory new. Protected by the housing, it remains as Chesapeake Gray and velvety to the touch as when it sat for sale out front of one of Raymond Loewy's freshly streamlined dealerships. Mark is getting set up to sandblast my truck's tire rims, but he needs silica for the blaster, so we make a Farm & Fleet run. It doesn't take two of us, but this is hardly the point. We pick up primer, black paint, rubber gloves, the silica, and two packs each of peanut M&Ms and Reese's Pieces. On the way home we pass a beautiful woman walking down the service road. I twist half out of my seat for a look. Then I look back at Mark. He's grinning. "A married man," he says, "learns to turn his eyes and not his head."

It's a good afternoon then, him blasting, me scrubbing. It takes a long time to do the grille. I have to work the pad in and out and up and down the vertical slots like I'm flossing the iron teeth of some robot. Now and then the pad catches an edge and the drill kicks back violently. When I finish, I take the sanding pad to the truck bed. So much time has passed since it was sandblasted that you can tell where we grabbed it to move it, because our salty handprints are revealed as rusty petroglyphs on the steel. It doesn't take much to touch them up, just a quick buff. Up above the retaining wall and behind a row of hostas, Kathleen has come out and is painting the house trim. She and Mark were able to buy this place on the cheap because the previous owners had rendered it unlivable. Among other things there was an abundance of cats, and drunken revelers had pissed down the basement steps. They've poured sweat equity into the place and now it's shaping up, but there's still a ways to go. I wonder how my sister feels about me roping Mark into this project, sucking up so much of his time. But I guess mostly we just enjoy the chance to spend some time together. I enjoy grown-up banter with my sister, because, thanks to the permanence of certain memories, I am always surprised that the little blond toddler who was paddling around in footie pajamas when I left for college has become this woman. Sidrock is in his crib beside her, chewing on an oak leaf and drooling on his toys. Mark and Kathleen were married in one of those speedy Las Vegas wedding chapels. "Same one as Demi Moore," Kathleen told me. They bought the videotape, and I watched it once. Kathleen was radiant and Mark stood back in a tux, his fingers twitching as though they were unused to hanging empty.

Despite my backsliding in the areas of tears and rage, it is my conviction that over the past several decades, the repression of feelings has been undervalued. After a lifetime of being harangued to *let it all out,* I am heartened by recent studies indicating the people who repress their emotions have a higher heart attack survival rate than people who are overtly emotional. I know people who are *constantly* "letting it all out," and their spirits remain consistently unimproved. I humbly submit that the world could do with a little more keeping it in. Sometimes caring people tell me I am repressing my anger. My chosen response is to meet their gaze intently, let one eye drift slowly inward, and reply: *"Yes. I. Am."*

And yet, compared to Mark, I am Richard Simmons. Mark is an eighth-level Zen master of stoicism. His philosophy can be distilled to three words: "Walk it off." He uses it in every context. Hit your head on the hood? "Walk it off." Burn your hand on the exhaust? "Walk it off." Wife left you for the Schwan's man? "Walk it off."

He says it all the time. He says it when Kathleen spills the paint. He says it when Sidrock raps himself in the head with his bottle. He says it when he hears one of his coworkers complaining about overtime.

He means it.

It works. Although to be fair, Sidrock isn't walking yet.

With the school year starting up again, Anneliese isn't able to wander up as often and stay as long as she has during the summer. We've also questioned lately whether we are cut out for some more standard long-term living arrangement. There are no histrionics, just long talks and quiet thinking that leaves us, as Anneliese put it, with "butterflies, buttermilk, and vinegar" in the gut.

We've been short of rain, and the garden has waned some again, but I manage to put away a few things for the winter. Mom gave me a big clutch of thyme and the tomatoes are coming in decent now, so I get out my foil roasting pans to make paste and stock the way I read in *Think Like a Chef*. I line the pans with sprigs of thyme, then pack in the halved tomatoes and a handful of unshucked garlic cloves, sprinkle in some sea salt and fresh ground pepper, drizzle on the olive oil, then slide the whole works in the oven to break down and mingle.

In between ladling off the tomato stock, I pack up some chicken breasts. First I lay out rectangles of tin foil and make a little bed of thyme in each. I settle the breast on the thyme and pack it with diced onions. I pulled the onions from the garden this morning and they are frankly sad, about the size of Ping-Pong balls, but they are my onions, and I will use them. Before I seal the foil, I add some soy sauce. After four breasts, I run out of onions and switch to making lemon chicken. The lemon balm has come on well, so I use the fattest leaves to line the foil. Then after topping the breast with a lemon slice, I wrap the whole works in leaves. Add salt, pepper, olive oil, and some capers, and seal the foil. I put all the breasts on a flat pan in the freezer. Later when they are frozen, I will vacuum pack them with my sealer, the one that makes the little farty engine noise when you push the button, As Seen on TV! Before sealing, I like to hold my index finger up and declare in a tone of wonder, "Just one touch!" If you try to seal the packages before they're frozen, the juice gets sucked out of the tinfoil. It's a baroque, work-making way to go about things, I guess, but it beats TV dinners, and you feel good to be stowing some things away against the coming winter.

When Anneliese calls and says she and Amy can run up for supper, I retrieve three of the packets from the freezer, where they've only half-frozen. Later while we bump around the kitchen rattling dishes and making fresh salad as the chicken bakes, it's Greg Brown on the CD player again, singing, *"where the kitchen is happy, love has a chance . . ."* and we make a note of it.

August has always been my month of resolutions. Forget January and the artificial premise of the New Year, it seems always to be August when I resolve that next year I will pare down, clear the calendar, and focus on doing a few simple things and doing them fully. This year I find myself entertaining visions of a one-room shack with one table, one chair, one skillet, one potbellied stove, and a wood-splitter's ax sunk at an angle in the doorjamb. Menu: mainly biscuits and bacon. Occasionally, just to test my mettle, squirrel. For dessert, apple pan dowdy. Certainly a measure of this reactionary navel-noodling can be attributed to the standard metaphorical casting of autumn as the season when winter's deathly breath first fogs your rose-colored glasses, but on a more fundamental level I think it has to do with the reaping of gardens and good intentions, both of which tend to come in well below spring's predictions.

CHAPTER 10

SEPTEMBER

I WAS RAISED BY strong women. Of course they could only do so much. I use the term *raised* in the perpetual sense, because the work continues. There is my mother, of course, sentenced to nature's most blessed curse, in which the female is expected to give of her body and blood in the rearing of a creature bound to bring trouble on the house. Not to mention the heart. A child is prayer and worry wrapped in a blanket. Tax deductible, yes, but oh, the hidden costs. You might describe my mom as the valedictorian homecoming queen who wound up a God-fearing homemade granola–slinging Florence Nightingale in a maxiskirt and construction boots stuck on a cow farm. Over the years she has taken responsibility for the care and feeding of legions of children—some conceived, some adopted, some fostered, some delivered by the county for the weekend, others for a lifetime. She is slight of build, and (to use her phrase) *just mortified* by public attention (thus I write of her in the broadest terms), but I have watched three firefighters rush to her with an unconscious baby and then enclose her in a semicircle of hulking apprehension while she calmly gets the kid breathing again. I have also seen her up to her elbow in the rear end of a sheep and giving rescue breaths to a newborn Holstein calf. (Mind you, not simultaneously.) For forty years she has raised a constantly fluctuating passel of tots, drawing on her wits, fifty-pound bags of oatmeal, and a fistful of coupons the size of a bad UNO hand. There were undoubtedly sleepless nights, but she never betrayed them.

For balance, I should also tell you she has imparted certain ineffi-
ciencies and weirdness, including an inability to focus in the midterm
(short-term emergencies—*like a laser;* long-term dedication to principle
and task—*can do;* finish one load of clothes without being distracted by
a mildewed *Reader's Digest,* the song of a wren, or the knowledge that
the mail has arrived—*not so much*); a propensity for impromptu flop
sweats (a familywide trait—my aunt Pam holds several world records);
and a habit of flicking her hands against her legs when she is discomfited.
Both my brother John and I have inherited versions of this last tic, effectively killing our chances of achieving success in the world of high-stakes
poker, international espionage, or, for that matter, dentistry.

My Grandma Peterson lived in fourteen different homes before she was
eighteen. Then, after a brief marriage that didn't work out, she moved to
a different town and married my grandfather. Not counting half a year
spent living in motels, Grandpa moved her in and out of another twelve
homes. Over this time she raised five children of her own and took in another twenty-eight foster children. She did her baking with a .22 rifle at
hand and was known to step away from the stove to snipe feral cats and
once an incautious woodchuck. My deerslaying uncle used to wax a bit
windy about his abilities as a marksman until the day Grandma took up
arms out behind the barn and schooled him in a manner he will to this
day not discuss, although one hopes the perforated targets have been
retained as evidence. In my favorite snapshot of her, she is leaning over
Dad's woodpile, detonating my brother's five-foot-long black powder
muzzleloader. The air is hung with smoke and powder sparks are still
arcing downward. I should add that she was scrupulously honest, deeply
compassionate, and jumped rope until the age of at least seventy-two.

Grandma Perry burned at least one pack of Carltons a day, and on
tough days knocked back a martini before supper. I recall her sticking
her wadded Clark's Teaberry gum on the plasticine of the cigarette pack
before lighting up, and my brother recently reminded me of the way she
held the handle of her fishing reel between her thumb and third finger
so as to free up the index and middle finger for the ever-present heater.
Sawed off and short-tempered, she favored sayings like "well hell-up-

a-tree" and "that woman could knock a bulldog off a gut-wagon," and she once declared that a certain parsimonious fellow was "tighter than a gnat's ass around a rain barrel," but the phrase she held closest to her heart still remains on the masthead of the local humane association she helped found: *a voice for those who cannot speak.*" She was fierce on behalf of animals, from picking up local strays in her orange Duster to serving as president of the state humane society. She had a way with human strays as well. The animal shelter employed its fair share of down-and-outers, and Grandma knew her way to the bail window of the county jail. I was present more than once when someone knocked on the door to reimburse her for covering their bond. She wrote heartfelt poetry, admired Albert Schweitzer, and spent her early married life alone, raising my father on her own with Grandpa gone to the shores of Iwo Jima.

Remaining within the family tree, there is my tiny aunt Sal, so bow-legged she couldn't catch a pig in an alley, but sharp enough to leave the smoothest horse traders this side of the South Dakota Hippodrome weeping over their slit pockets. Aunt Mabel, eighty years old and currently charging off to every cultural event within a fifty-mile radius of Spooner, Wisconsin, usually with a carload of restless contemporaries. Aunt Meg, who took time out between hairstyling and running her own café to drive an eighteen-wheeler, hauling corn syrup and refrigerated goods across the country behind a gigantic black Freightliner. Aunt Annie, who got all the cows milked before heading to the hospital where she delivered baby number three directly upon arrival.

○ ○ ○ ○

Somewhere around last June or so, I was talking to Anneliese about how beautifully it seemed to be going with us, and she said she would be reserving her judgment until six months passed. That was a rolled-up newspaper to the snoot. Counting from our first date in mid-April, we have a month and a half to go. Anneliese wrote me a note recently saying she liked "how we work—deeply and easily." I, in turn, told her from the bottom of my heart that I was grateful for her "reasonable-ness." This is the complimentary equivalent of a vacuum cleaner for Christmas. And yet she confirms that reasonableness by consistently giving me do-overs in the wake of such dumbfounding tone deafness.

It hasn't been a complete skate. There are things. Little pop-up ghosts of the past, hints of disagreement over where we might live if we decide to make it official, our continued reasonable doubts about how either of us will manage the shift from long-term independence to peaceful co-habitation. We both have a desire to get off the grid, more or less. Keep the power line, but supplement it with a windmill and a few solar panels. Each of us hankers for a real garden and those chickens. Recently we took a ride to look at some land near where Anneliese was raised. We stood there under a gray sky, and Anneliese talked about the beauty of the place. She was right. It was beautiful. But when I stood there, I didn't feel a thing. Later I drove her a short way out of my village to another patch of land. Same situation, reversed. She couldn't feel what I felt. On the drive home I got grumpy and quiet and couldn't look her in the eye. We sat quietly for a long while after. Anneliese said she felt like the vinegar was back. Later, alone, I waited for the old hopelessness to return. It did not. Perhaps it was her clear blue eyes. Having failed at this time and time again, I hesitate to say, but I have never felt so placid.

I hope I'm right. Love is a contact sport of the heart. You can't take the hits like you used to.

○ ○ ○ ○

Beyond blood, I was schooled in powerful womanhood by the dignified Charlotte Carlson, who came to Chippewa County in a buckboard and lived to see the space shuttle launch. When my parents bought her farm, she was not bitter but rather became my mother's best friend and our surrogate grandmother. From her I carry an affection for potato lefse and sugar cookies and good whole wheat flatbread spread thickly with butter. There was Vernetta, the next-door neighbor who fed her hay crews thrice between noon and five o'clock, unloading hay wagons in be-tween. Nelda, the Wyoming rancher's wife who always wore a dress but could take out a prairie dog at two hundred yards (discounting the time she miscalculated while using her pickup truck hood as a rifle rest and put a straight crease through the sheet metal). Ramona, from the farm across the forty, who babysat us when I was young. She held my brother headfirst in the sink when I knocked him off the concrete steps with the screen door and blood poured from his brow. Thirty years later a woman visiting from somewhere else met Ramona and said, That Perry boy writes beautifully, and Ramona said, Well, that ain't the way he *talks*.

In 1984, I was admitted to the University of Wisconsin at Eau Claire School of Nursing and spent the next three years immersed in a world that was 3.5 percent men. To say my female classmates were my intellectual equals or better is neither conjecture nor condescension—it was a fact verified weekly in the form of coded test results posted on a wooden door. I consistently ran my finger down to the lower middle before locating my secret number. This had a way of rendering gender issues moot. My instructors were by and large fiercely intelligent vanguard feminists, eager to eradicate chauvinism in all its forms, although they didn't seem to mind asking me to help lift heavy patients, or strip to my shorts when it was time to demonstrate physical assessment or intramuscular injections. At the time, I got a little righteous about this contradiction. I have since come to understand blind spots are universal, and less a sign of inconsistency than proof of humanity. I propose this in part because I could sure use the wiggle room.

Sometimes when I am snuffling along nurturing my vision of Anneliese as the simple farmgirl of my cornball dreams, I will attempt to join her in the company of certain women and run face-first into an energy shield that causes me to lumber backward like the bull who lays his dumb wet nose on the electrified fence. I tend to put my faith in physics over mystery, but for all the farmwives and feminists and well-armed grandmas in my life, there have been a small convention of others who have modeled more ineffable powers. Galinda, the neighbor up the road who, in the process of allegedly unruffling my field, hovered her palm about a half inch above my right scapula and said, "I feel a lot of heat *here.*" I had been having spasms there for a month since getting hit in a football game, but there was no mark or bruise, and I hadn't said a word to Galinda. I remember thinking, *Well now.* There is Laurie, a woman who—I don't know any other way to put it—glows with invisible light.

With her shawls and sculpture work and barefootedness, she is the personification of an earth mother, but I have seen her rise up and put down
a strong man while the rest of us were content to avoid eye contact and
whisper behind our hands. Recently she told me she believes we are
all energy created by a purposeful magnetism and that the catalyst is
love. This may explain the invisible light. The first time Anneliese and
Laurie met, they hugged, and then later each sought me out separately
to remark on the other's aura.

Finally, there is my cousin Alice, a painter and poet and stained-glass
glazier, who claims certain extrasensory powers and has been known to
stumble over force fields. She says there are times when she is led by
birds. I am leery of such talk and would be dismissive, except that Alice
once outbid a stranger to buy back her deceased uncle's 1950 Dodge
pickup at auction, the sole reason being that she hates to cry, and I am
pleased by the idea of psychic plus stoic. Alice once told me artistry does
not reside in motivation but rather stems from showing up, with the
intent to be honest. When I meet a dreamer with calluses I try to shut up
and listen. And lest everything sound too glorious, I am also remembering Cerise, who taught me that a strong woman can wind up battered in
ways no weak man ever would.

⚙ ⚙ ⚙ ⚙

Around the middle of the month I get a chance to work on the truck
again. As I'm getting into the car I stop at one of the raised beds and pick
a green pepper, a cucumber, and some cherry tomatoes. The garden remains pretty much a flop (after a long dry spell we're getting rain today),
but even the most middling return on my efforts is rewarded when I bite
into that pepper. Eating it like an apple as I drive, I marvel as always at
the sweet watery crunch, one of my favorite "clean" flavors, reminiscent
as it is of fallen rain. Conversely, I find that when cooked, green peppers
become snotty and overbearing, capable of dominating and ruining the
spaghetti sauce.

The radiator is fixed, and I have it in the backseat of my car. After I
unload it and Mark inspects it, we start working on the fenders. Mark
will be able to cut and transfer patches from one of the cannibalized

fenders. For a few of the smaller holes outside the range of the patches, Mark has me hold a flat chunk of copper on the underside of the fender while he uses the welder to fill the hole with a molten bead. The bead doesn't cling to the copper. Later I go back and knock the dome off the bead with a grinder, then go back over it again to sand the bead flat. You're left with a smooth surface and a small line of suture where the fresh bead and old steel meet. When we've finished filling holes, I take a wire brush to the inside of the fenders. After I've scuffed away all the loose flakes and spurs of oxidized steel, I brush the entire surface with rust converter. It goes on foamy white, like spilled milk. As it soaks into the surface, the rust takes on a bluish tinge. Several hours later what was once rusty looks oily and dark, and the rust is inert.

The bottom edges of the fenders (including the ones we robbed off the junker) are badly eaten with rust. We toy with several options—patching the holes, welding a steel strip around the base of each fender—but finally Mark suggests he simply trim them back an inch or so. It seems like a simple enough approach, and the International fenders of the day hung like big steel drapes anyway, so I give him the go-ahead. Once again the purists weep, although I have been reading more about Raymond Loewy, and one of his pet phrases was "areas of examination"—that is to say, what can be tweaked to favorable effect? Loewy was famous for cutting chunks off his Cadillacs and BMWs to reconfigure their lines in a manner he found more pleasing, and I believe Mark is up to the task.

Kathleen calls us in for supper. She's made a Crock-Pot of sloppy joe barbecue. After eating and chatting, I head back out to the shop and work alone. My goal tonight is to pull the seats. The truck came standard with a bench seat, but I tore it out in the late 1980s for various reasons including the fact that the bench seat was very jouncy and in combination with the tanklike three-quarter-ton suspension tended to hurl you ceiling-ward at the slightest bump in the road. Having tired of the repetitive neck injuries and cranial knots, I decided to try something different. I had also recently begun working as an EMT, and having extricated my first few bad car crashes, I wanted to put in some seat belts.

I began the project by driving some distance to a junkyard somewhere north of Highway 8, where I detached two bucket seats and a pair of lap belts from a moribund Ford Maverick. I have absolutely no recall of why I took them from a Ford Maverick, but there you are. I proceeded back to my father's farm shop and commenced the renovation.

The bench seat came out easily enough. It was simply bolted to the floor, although it appeared someone had raised it by inserting two wooden blocks beneath the rails. I stuck it in my father's old chicken coop where as far as I know it remains to this day. I then drilled holes through the cab floor in a pattern matching those in the rails of the Maverick seats. I hoisted the first bucket seat into place behind the steering wheel, bolted it down, and hopped in to judge the feel. My eyes were looking directly at the horn button. I had to reach for the steering wheel like I was hanging laundry.

Disappointed but not deterred, I unbolted the seat, pulled it out, and went around behind the shop to my father's scrap-iron rack, where I retrieved a mismatched bundle of angle and strap-iron. A flurry of hacksawing, grinding, blowtorching, drilling, welding, and bolting followed. When it was over, I had created a steel frame to be bolted between the cab floor and the bucket seat to provide the necessary elevation. Furthermore, it gave me an anchor to which I could bolt the lap belts.

The unpainted frame was esthetically iffy and weighed nearly as much as the seat itself, but it worked. This time when I climbed into the cab, I could see over the dash quite nicely.

The trouble with so many of these spur-of-the-moment male projects is that you run out of time. It was growing dark and I had to be back at work in the city that next day. So for the next several months I tooled around in the truck with nothing but a large cavernous space where the passenger seat should be. I was dating at the time and the truck was my only transportation. The woman I was seeing had already endured me for years, we had been off and on since high school. She had been tooled around in my dad's farm truck, and she was the girl who wound up with her face wedged between the windshield and the dash when I borrowed my grandfather's sedan and tried to shift the automatic transmission by tromping the automatic brake like it was a clutch. I once took her snowmobiling and failed to notice she had fallen off until I was two forties away.

Running her around in that seatless truck, I didn't want to be an insensitive boob.

So I put a bean bag chair over there.

She did eventually marry a man *named* Mike.

I soaked the bolts down with WD-40 beforehand, but they aren't budging, so I get the sidewinder grinder and shear them off with that. I wear earplugs and goggles, but the sparks that fly off the grinder and ricochet off the surrounding steel land on my arms and the tender antecubital space of my elbow, atop my balding head, and even my eyelids. It becomes a sort of miniature G. Gordon Liddy test to ignore the pain of the iron tinder as it hits the skin, to stand without flinching. These are little bits of molten steel, but despite the sting, they cool before burning into the skin, so it's a way to work on your focus. The grinder is also giving off fine plumes of powdered rust that settle fuzzily on the hair of my forearms, on my ears, and scratch in my eyes.

When the last bolt is severed I tip the passenger seat off the bolt stubs and drag it, homemade frame and all, to the center of the shop, where I study it awhile, suddenly struck by the fact that I have no clear memory of who I was when I put that seat in. It's one of those little markers in time that trigger a cascade of other memories—many of them scintillating in their detail—and yet leaves us with one gigantic gap: Who *were* we then? You go back and examine your life and it's like unfolding one of those segmented drawings where each person draws a part of the body—the feet, the knees, the waist, and so on, up to the head—without seeing the rest. When you unfold it, it's funny to see how things match up, or don't.

I work late, well after midnight. At one point, having had enough country and oldies, I rummage through Mark's stack of CDs and pull out the Nirvana *Unplugged in New York*. I missed all but the peripheral aspects of the grunge movement, although a friend recently took one look at the way I was dressed and said, "Seattle, 1989." When Kurt Cobain was ascendant, I was still transitioning out of hair metal and exploring the New Traditionalists of country music. So I always listen to Nirvana with the sort of befuddled appreciation that, indeed, something was going on there but I should just enjoy what I can and not insult the involved parties by pretending to get it. This brand of dispassion is one of the privileges of aging. As the decade-old concert unfolds, I putter away, pounding out the bolt stubs (they have rust-welded themselves to the underside of the cab), painting rust converter on the back fenders, cleaning up some. The last song on the album is a cover of the age-old folk ballad, "Where Did You Sleep Last Night." Written sometime around 1870 and known variously as "In the Pines," "Black Girl," and "Black Gal," the song was made most popular in the 1940s by Lead Belly. Like many folk songs, it has permuted through all the hand-me-downs, but its elemental darkness remains stark. Cold wind, dark pines, decapitation by train, these are not images prone to ameliorate over time. *"Girl, where did you sleep last night?"* is a question that delivers a powerful jolt to the liver no matter your age or era. Throughout the early verses, Cobain's voice conveys a malevolent desolation that finally explodes in a harrowing squall so blistering it seems you could turn the boom box toward the truck fenders and watch the paint bubble. The grit on my face, the rust at the back of my mouth, the bits of grinder slag peppering my arms, it seems he was, holding them all in his throat. The guitar chops its way to the end of the song, the audience cheers and fades, and I am left with nothing but the hum of the fluorescents.

It's 1:30 A.M. when I drive away from Mark's shop. On my way home I pass through the village of Cameron. As I approach the town's solitary stoplight, there are no other cars in sight. I am thinking about Anneliese and I am sitting at the light for a good time before I realize I've pulled to a dead stop on green. I accelerate and pull away, checking to see who might have noticed. I'm half a block from the intersection when the village police car pulls into my rearview. I'm doing 35 in the 35, no worries, but he follows me out of town, past the high school, past the exit to the freeway, and just as I begin to accelerate for the 55-mile-per-hour zone, his lights come on.

After a long wait while he runs my plates, the officer exits his car and approaches. He sweeps the interior of my car with his light and I cringe, because the car is swimming in junk and truck parts and road food garbage. Then he puts the light on me, and of course what he sees is a dirty, unshaven nut job out driving erratically after midnight. "Good evening, sir," he says, lowering his flashlight beam just enough so that I can see him. He is a small fellow, and disturbingly young. His gun belt hangs on him like he dug it out of Daddy's drawer to play dress-up. Leaning down to speak, he also tries to catch a whiff of my breath.

"Have you been drinking, sir?"

"Nope. Not for thirty-eight years."

"Well, sir, I noticed you drifting over the fog line several times."

This is flatly bullfeathers. If he had said he became suspicious when I spent five minutes camped at a green light, I'd have been down with that. But here he is plainly fishing.

"May I see your license, sir?"

With an eye toward his youth and his gun, I explain that I have to dig around some and wait for his permission. While I'm digging I'm thinking he's going to love my license, which features a photo of me with frayed butt-length hair and an overgrown beard. It could be a membership card for the National Association of Deranged Street Prophets.

He spends a long time back there in the squad, allowing me time for reflection. I have always believed that good cops can't be thanked enough for doing their impossible job, and as far as I know, the toddler back there running my plates is one of the good guys, but I'm surprised at how every time he spoke to me I had this urge to turn him over my knee. My reflection yields no epiphany beyond the fact that one becomes an old coot by increments, and here's one now. And here he is back at my door, handing my license through the window. I can go, he says. Then he leans back in the window.

"But, sir?"

"Yes?"

"Try to pay a little better attention to your driving."

"Ooookayyy," I say, turning red in the dark and wishing I had the guts to add, ". . . *Spanky*."

The remainder of my drive was consumed with muttering.

○ ○ ○ ○

In 1995, Natalie Merchant released a music video (for the song "Wonder") populated with beautiful women of all ages and persuasions. Beautiful women of course are a staple of the genre, but these women were different. Tawny Kitaen unfurling herself across the hood of a Jaguar is an incitation; the women in Merchant's video were a revelation. They were beautiful through a sublime range of age, form, and physiognomy; the cumulative result spoke to the beauty not of women, but of womanhood. I saw the video only once or twice, but it made a powerful impression, and I was reminded of it again when I arrived at Anneliese's house today to find six women (counting Amy) in the dining room. Here was Anneliese not in the context of the usual self-centered *me* but in the context of the strong women in *her* life. Her mother, Donna, a woman who has always made me feel welcome but never entitled, and right she is; Jaci, a former teaching colleague and the person who must be given credit for Amy's Seabiscuit phase; Bibi, another teaching colleague from Colombia who helps Amy with her Spanish; and Heather, a dedicated missionary who came to Anneliese's aid in the difficult early weeks after Amy was born.

It's so easy to get so caught up in our brief little history since the library reading in Fall Creek. You forget sometimes what a disruption you are. And how late you have entered the game. After our trip to Colorado, I talked with Anneliese about where she turned for help in the time surrounding Amy's birth. Her mother came, of course, and there was Katrina, the local church minister. Stacy, the woman with the doctorate in physiology who stood in as birthing coach. The midwife. The doula, a woman I may never meet whose role as I understand it was to operate somewhere between coach and midwife. Jen, the friend who accompanied Anneliese on a journey to Guatemala after the pregnancy became known. And, in a nod to the men, Ryan, a friend of the group who paced the halls during the birth, which turned dangerously difficult, with Anneliese in surgery and too weak to stand for days.

But all those fierce women. When I see a few of them around the table today, I don't feel I've missed anything, but rather that I am being allowed something and had better pay attention. When Anneliese came home from the hospital, there were these women to help her. But it was she who did the feedings, she who went back to work, and she who prepared and delivered the defense of her master's thesis on poetry as a form of shamanism based on the Guatemalan poet Humberto Ak'abal. Cumulatively this reminds me for the forty-seventh time that I shouldn't do all the talking.

I've read the thesis with its references to innate purpose and hexagonal shapes as an image of humans as the center between worlds, and after balancing it with images of Anneliese still in high school, bucking hay bales for the neighbors, or feeding her stepfather's cows, or handpicking another truckload of his onions back in what would become difficult days and her mom would move away, and I am forewarned that I am not in the company of someone who feels the need to hang on my every word. When she still had Amy in her belly she climbed a Guatemalan volcano in the dark and stared at the lava below. Sometimes it is good for me to look at the two of them together and think of that.

A little voice inside me is saying the man who sets out to celebrate womanhood and its constituency is waltzing through a minefield set in quicksand. Good intentions lead to woman-warrior overrevving or patriarchal head-patting. Contradictory elements abound. The man moved by a Natalie Merchant video to consider all womanhood is by no means inoculated against the booty of Sir Mix-a-Lot. Anneliese captures me with her character and spirit, but I am not blind to the way she walks. Having gallumphed the streets of Manhattan and felt my heart stricken every six feet based on nothing more than mystery and outward appearance, it seems carnality is ridiculous and essential. Rare and wearisome is the fully emancipated man. Our only hope is to be judged on the balance of our actions. Sometimes the power of a woman is no more ineffable than a mallet. Shortly after my first date with Anneliese, a woman who had come to know us both in separate circumstance sent me a firm note that concluded with a time-honored blessing: *Try not to screw it up!!!*

Once I was in a moderately fancy lakeside restaurant dining with a woman who to this day can post reasonable doubts about my character when a couple in a simple aluminum fishing boat docked and took a table. They were both middle-aged, both wearing discount store tennies and appliqué sweatshirts (an eagle on hers, a buck deer on his), both a little chunky, and both sunburned in a manner that suggested this was their one week off and, by God, they were gonna fish. They studied their menus in silence, and when the waitress came by, they spoke so softly and returned the menus so meekly it was as if they had not ordered but asked permission. From their dress to their demeanor, they were not one whit demonstrative. Then the woman turned for her purse and wiped out her water glass. The clatter and splash cut through all the jabber and you had that pause in the action when all heads turn, then the rhythm of the room resumed. The woman stared at her place mat, her sunburn heightened by a red flush. And right then that man leaned over and put one arm around her shoulders and gave her the softest little kiss on the cheek.

I have been trying to live up to that man ever since.

On my next trip to the shop, the weather has gone cold, and the building has the feel of a clubhouse again, muffled in warmth. I do a lot of miscellaneous things: put nuts on the taillight studs, clamp the taillight brackets to the bed so Mark can weld them in place, take a wire brush to the floor of the cab to prepare it for repainting. The speedometer has never worked, so I slide under the truck on the creeper and detach the cable where it inserts in the transmission. By fiddling with it I can get the odometer to spin, but the speedometer needle doesn't budge. Intending to rob the speedometer off the L-180, I find that the indicator needle on that unit is missing completely. Honestly, what good is a speedometer in an International? If I get caught speeding I should get a little plaque or something. Before I leave Mark comes in and we lift the detached bed into place on the frame. He finished trimming the fenders last week, and the truck looks more nimble. It's not, of course, but it looks good. "The entirely new fender line," wrote Raymond Loewy after streamlining his 1943 Caddy, "visually seemed to lengthen the body and provide a feeling of speed." Yeah, buddy.

Mark says he's about ready to paint the thing. We're getting there.

○ ○ ○ ○

I have had the opportunity to watch the women of my family move through the days of their children dying, and within this circumstance beyond all others seems to lie the very paradoxical essence of womanhood. That fierce and mysterious capacity for life, hamstrung by its one great vulnerability: love. To see a strong woman living beyond the death of her child is to see all women living and grieving in this man's world. One night while driving in the rain far from home and after midnight, I was listening to Patty Griffin sing her live version of "Mary," a song she wrote for her grandmother. In the chorus, Jesus kisses his mother, tells her he cannot stay, and then, as he goes flying to the heavens, Griffin's voice soars into the line, *the angels are singing his praises in a blaze of glory.*" Griffin draws her voice inward now, breathing out the final line with weariness and resolve:

Mary stays behind and starts cleanin' up the place . . .

It is the history of womankind in a single line. I flashed on images of my mother and my grandmothers beside caskets holding their children and sometimes their men. I pulled over for a while, played the song twice more, and hoped the tears on my cheeks would count for reverence.

CHAPTER 11

OCTOBER

IT WAS RECENTLY my duty to describe my left testicle to a strange woman over the telephone, a privilege for which some men would pay upward of $3.95 a minute, but which I found discomfiting, although markedly less so than the moment when I sat in a paneled office beneath a stuffed deer head and described that same testicle to my insurance agent, Stan. Talk about your festival of averted gazes. Stan bent to his paperwork with all the diligence of a first grader determined to win a blue ribbon in penmanship, crossing and recrossing his *t*'s, carefully scribing each loop and line, no doubt desperate to get everything right the first time, terrified he might have to repeat this little sharing session of ours. I felt bad for him. There was a palpable sense of him yearning to talk fender benders. When we moved on to discuss the blind spot in my left eye, he sagged with relief.

The deal is, the damage is adding up. Here on the cusp of forty, I am daily grateful for my health and do not for one minute request special pleading. Still: *Things fall apart; the centre cannot hold.* One hesitates to sully Yeats with equations to my nascent gout, but with each little hitch and failure, the message is: you are in possession of a machine programmed to self-destruct. I used to have weird dreams about my teeth coming loose; now I have weird dreams about my crowns coming loose. In addition to the eye and the testicle (specifically, its benign epididymal cyst, otherwise known as a *lump*), my inventory includes sand in the gears of my neck (I find myself checking blind spots in traffic by pulling half a chin-up on the steering wheel and then rotating my cranium, neck and shoulders as a single unit) (very little-old-man), a frayed rotator cuff, persistent tinnitus, hyperacusis, a world-record kidney stone, transient numbness of the left leg, a partially detached clavicle, a little click in my thumb that has lingered since I jammed it in someone's shoulder pads during a Friday night football game in roughly 1982, and yes, here lately, in both big toes but particularly the right, the first twinges of uric acid accumulation.

I have been purchasing my own health insurance since 1992, and for years I put my faith in luck and youth and covered myself with a high-deductible catastrophic medical-surgical policy. When my left eye went goofy in 2000, the tests and treatments left my savings pretty much tapped, and I decided it was time to upgrade. I quickly discovered that my faulty left eye and irregular left testicle made potential insurers scarce. One company simply ignored my inquiries. Another rejected me outright. I remember when I read that letter, I got a cold little clench in my gut. I imagined my eye growing darker, or my liver failing, or some red-light-running yahoo taking a bead on my femur. Based on extrapolations from the invoices for my eye, I did the math and got this vision of my sturdy old house, my savings, and my used Chevy spinning down a drain.

Stan had handled my car insurance for years, so I made an appointment and told him of my troubles. He submitted an application to a Major Insurance Company and shepherded it through, which led to my telephone conversation with the nice lady. A few days after that heartfelt exchange, Stan called with good news. I had been approved by the Major Insurance Company. Mostly. "You'll have to sign a rider," said Stan, quietly. "One for your eye, and . . . and . . . and one for . . ."

"I understand," I said. Poor guy. I drove to his office and signed the paperwork, including two "Special Exception Rider" forms that drastically limited coverage of my left eye and any complications related to the cyst. Having done their own math, the Major Insurance Company wanted no part of those parts of me. I was surprised at how easy it was to make the trade-off. During my health insurance search, I read consumer literature that said you should never accept a policy with riders, but I was ready to deal. A friend told me that when it comes right down to it, you'd give your left nut to have some health insurance. On a related matter, should you ever join me in a bar brawl, you will note that I lead with my right.

○ ○ ○ ○

My truck and my garden will languish this month, because I am on a book tour that has me on the road all but two days in October. It helps to know that we got such a freeze on the first day of the month that the cucumbers are done for good. I will miss duck hunting and rutabaga season and six or seven lamentable metaphors spawned by the falling leaves, but book tour is an all-expenses-paid scavenger hunt in which you run around the country attempting to collect the items on your list: bookstores, radio stations, public access television stations in the back of tire shops, hotel rooms, rental car return lots, departure gates, coffee, small towns in Michigan. Your life boils down to showing up to yap. But I am well taken care of, the people I meet run to the high end of well-read and pleasant, and I collect miscellaneous anecdotes. A literary escort in Kentucky once told me she knew she had moved to a small town when the editor of the community weekly—unable to send a photographer to cover a local wedding—simply dropped the newspaper camera off at the reception and asked the family to return it with a few decent shots. She added that soon after her arrival in that same town, she had car trouble. A man strolled over and helped her get the vehicle running. She took his name, gave him twenty bucks, and wrote a letter to the local newspaper lauding his sterling character. In the letter she said she assumed that this man's actions spoke to the character of the town in general, and that as a newcomer, she was thrilled to be part of such a community and would do her best to maintain the standard of conduct. Her letter was published on the back page. On the front page was an article describing how the man in question had taken the twenty bucks, bought beer, got drunk, and stole a truck.

The term *literary escort* has always struck me as sounding simultaneously high-brow and naughty. Synonyms for *literary escort* cumulatively include wrangler, restaurant critic, psychiatrist, race car driver, smooth

operator, expediter, ego polisher, brilliant conversationalist, silent partner, procurement artist, navigator, parking savant, and restroom locater.
Plus they fill a seat at readings. I spend the majority of my book tours running solo, during which time I am perpetually sweaty, late, lost, nervous,
tired, lost, late, and furthermore sweaty. But occasionally if I am in a big
city or at a large event, the publisher arranges an escort, and suddenly
book tour becomes a day spa. A good literary escort shaves fifteen points
off your systolic blood pressure. Above all, they know where everything
is. I can spend a full day hammering around San Francisco and make
maybe one radio interview and two bookstores. With a literary escort, I
just sit in the passenger seat and get delivered. We go from bookstore
to bookstore without pause. The best literary escorts have the same sort
of ineffable cool I usually associate with old-school television gumshoes.
Once in Chicago I was being taken to a downtown radio station by a
legendary escort I shall call Bill. There wasn't a parking spot in sight. Bill
eased up to the curb of a major high-tone hotel. This was clearly a No
Parking zone. As we exited the car an immaculately uniformed doorman
approached. I got the usual law-abiding flop sweats. Bill strode ahead of
me, and as the two men passed, their hands met briefly. The doorman
gave Bill a collegial nod, and we were on our way. And to this day Bill
remains one of the coolest guys I know, because midway through some
story in some bar somewhere, Bill can lower his beer glass coolly and say,
in all truth and without batting an eye:

   "So I just slipped the guy a sawbuck . . ."
   I get shivers.

I call Anneliese nearly every night, although most times we keep the
conversation short. We are developing a counterintuitive theory that
long phone talks during periods of separation tend to fray more than
they bind. You have two people, both executing the responsibilities of
the day, each quite understandably focused on the view from their respective perspective, and if the conversation goes on too long, a subtle
one-upmanship emerges in which each person attempts to prove to the
other that life on this end of the phone is no bed of begonias, either,
and before you know it, long-distance pillow talk becomes an interrup-

tive and blameful downer. Perhaps one day I will recant, but for now I
vote for short phone calls. I do try to drop a sweet note in the mail here
and there. Raised by a sometimes-single mother who climbed poles and
strung telephone wire for a living, and having been a single mom herself
for three years now, Anneliese is not given to pining over the tenor of
my angst as I gaze out a hotel room overlooking San Francisco or order
room service in Albany.

But gosh, it lifts my heart to hear her voice.

Three years ago this month, I was walking across a plowed field beneath
clear skies when a shadow fell across my left eye. At first I thought it was
a prodigious floater. I blinked hard and rolled my eyes, but the shadow
remained. There were no symptoms—no dizziness, no headache—so I
chalked it up to some trick of light and continued walking. Half an hour
later, while trying to read a book while perched in a deer blind, the dark
shape kept sliding on and off the page. It eased through the periphery of
my vision, only to dart away when I tried to pin it. I could study it best
when I gazed up and off to the left, in the manner of a boy caught looking at a pretty girl at the coffee shop. After some time, I could make out
that the black spot had a shape: a fuzzy-edged delta of darkness, growing
from a narrow point, then widening in a slight curve before squaring off
at the edge of my visual field. The wedge shape made me nervous. It
reminded me of the visual field cuts I saw diagrammed during nursing
classes on stroke. I hiked out of the woods at sunset. It was a Sunday, so
I called a friend who worked in a hospital emergency room. She in turn
referred me to an ophthalmologist acquaintance. He was carving jack-o-
lanterns for his grandchildren when I called, but said he would see me.

It was contextually otherworldly to follow Dr. Olson into the dark,
empty clinic as he flipped the lights on and led me down the hall into an
examining room. I was worried about my eye, but found the examination process strangely comforting. For one thing, in times of uncertainty
we always like to place ourselves in the hands of someone with knowledge, and Dr. Olson is trim and apt and conversational. He seated me
in a comfortable chair with an adjustable footrest. We did the standard

vision test, the letters on the wall, which always transports me to grade school when they test your eyes. I passed. He had me stare at a copy of the Amsler Recording Chart, which is basically on the order of graph paper. By closing my right eye and staring at the chart with my left, I could trace the darkened area with my finger to give Dr. Olson an idea of the shape and extent of the affected area. He gave me my own copy of the chart to take home and told me I should check over the next few days to make sure the shape wasn't growing. Then he put drops in my affected eye and killed all but the ambient light. While we waited for my iris to dilate, we talked about Wisconsin Badgers football games and Manhattan and pumpkins. The mood was relaxed and intimate, like a candlelight dinner. When it came time for him to peer inside my eye, the world narrowed down to the BB-sized point of light burning in the aperture of the ophthalmoscope, and all of it at the center of the still, silent clinic. Everything but the light was muted. There was something of peace and clarity.

Dr. Olson said there wasn't much to see in there, just a tiny whitish spot at the confluence of an artery no bigger than a thread. A lesion, he called it. The good news was there was no bleeding, no reason for me to be rushed off to surgery, no sign of a tumor or otherwise. The bad news, which Dr. Olson gave me straight up, was that the black patch was likely there for good. There was a shot that peripheral circulation might restore some of the vision over time, but this was unlikely. He would want me to come in tomorrow during regular hours so he could do a full battery of tests, but for now I could go home.

○ ○ ○ ○

At some level, all one-eyed people are mystic, if for no other reason than they quite literally see things differently than the general mass of us. Jim Harrison lost the sight in his left eye at age seven when a little girl stabbed him with a bottle, or that's one version. The blind eye figures regularly but not obtrusively in his novels, plays, and essays. I came to Harrison embarrassingly late, first reading him sometime around 1995. Every day since, I have tuned to his key and tried to write a line he might find credible. Relevantly or not, I find the best stuff tends to come sometime around two or three in the morning, when my left eye—lazy for years—droops closed and I just leave it shut.

Perhaps as a result of living one stick-poke removed from total blindness, monocular characters tend—in conversation and comportment—to convey an equanimous blend of fatalism and exuberance. I own a little patch of land next to a farmer and part-time trucker named Jerome. Jerome lost the sight in his left eye the year he turned sixty. He's getting treatments to restore full vision, but they have been ineffective and the eye is paining him. "I told the doc, 'Just yank'er out!' " says Jerome, slapping a pair of pliers resting in a worn leather holster he keeps threaded on his belt. He moves with the barrel-chested stiffness of a man who has spent decades bulling his way through hard luck, boulder-studded fields, plummeting milk prices, and *damn government shysters*, all the while keeping things together with a pair of pliers. I've known Jerome for a while now, and I believe him about yanking the eye, because last year he swiveled a quarter turn to get his good one on me and then shared his creed: "Mike, I pay my bills and treat people square . . . but now and then you gotta do something just to keep'em wondering."

Hank Carhart would agree. Hank is one of those small, lean, roughneck-looking kinds of guys whose two main possessions are a motorcycle and a dog. You see him and you think *bar fight*. Hank is a one-eyed land surveyor. There's a good joke there somewhere. When he was a kid the neighbor kid shot him in the right eye with a BB gun. "Just like Mom told ya, 'Be careful with that thing, you'll put somebody's eye out!' " he says, blowing cigarette smoke and laughing at his own joke. "It was like that Christmas movie where the kid catches one in the glasses . . . I peeked out from behind the tree and took one in the looker." He quickly adds, "I was shootin' at him, too, so it was legit." Code of the West.

On the advice of doctors, Hank kept the sightless eyeball in its socket, but by his mid-thirties it was giving him nonstop headaches, so he had it plucked. "Enucleation," he says. "I asked my doctor what we were lookin' at for a time frame, and he said, 'Well, I can't get you in today,' like it was a *haircut,* man!" After the eye was out, Hank wore an eye patch, but not for long. "It attracted freaky chicks," he says. "Freaky chicks and freaky people in general. The patch is a total weirdo magnet." Once a complete stranger asked for a peek. "I told him, man, you're the kind of sick fuck who would come up to a guy with a wooden leg and ask him to dance!" These days Hank wears tinted glasses ("Mostly to protect my good eye") and lets people gawk. As for myself, I once bought a patch at Walgreens, intending to wear it while writing and thus strengthen my lazy eye, but I have not been stringent and will be lucky if I can find the patch come Halloween or freaky chicks.

I intend, by the way, that the one-eyed category include the esotropic. It is the great curse of cross-eyed people that a convergent gaze tends to serve as the international symbol of comedy. Arraigned with Jerry Lewis I will plead guilty on that count. If, in a moment of levity, I am moved to pull a face, I tend in nearly all circumstances to stick out my tongue and cross my eyes. Perhaps cavemen did the same. What a burden it must be for a cross-eyed man to project solemnity. As an aside, I must confide I have long harbored an absurd jones for slightly cross-eyed women. I find the look vulnerable and specifically foxy. Think Karen Black and Barbra Streisand, circa 1970s (circa 1970s being pretty much implied by my use of the term *foxy*). I once dated a woman who went faintly cross-eyed whenever I leaned in for a kiss. My heart would melt. Until this very moment, it never occurred to me that the reaction may have been symptomatic of my presence or the angle of attack. I would never attempt to kiss my neighbor Bob the One-Eyed Beagle, but I have studied him closely and believe he has special powers, if only because he is a cross-eyed butcher with ten fingers. That, and he regularly says things like "That woman was hotter than bumblebees in a tuna can," which qualifies him as mystic in my book.

◇ ◇ ◇ ◇

The morning following my initial exam with Dr. Olson, I returned to the clinic for a complete workup. On the drive down, I kept chasing the shadow across the windshield. It wasn't your standard pie-shaped wedge. It had more of a wave to it. Kind of a fat longhorn shape.

Unlike the secret hideout of last night, I arrived to a well-lit waiting room filled with people. Most of the patients were thirty years older than I. I reported in, sat down with a magazine, and got immediately bugged by the shadow skittering back and forth on the page. I was experiencing the standard feelings of fragility and petulance precipitated when life taps our knuckles with the hard wooden ruler of mortality. Evidence of our impermanence is available on a moment-to-moment basis, and yet we constantly deny it. Two days ago, on Saturday, in that same deer stand, I had looked up from my book into the gray hardwoods and was struck by the astounding passivity of vision, how all the available light— even when presented in the limited chromatics supplied by a grove of leafless trees—floods through our cornea to be transduced and painted on the brain. I contemplated this and made some notes toward an essay on the wonders of vision, then discarded the idea as trite. It was a dose for both the romantic and the ironist in me when less than twenty-four hours later, that vision became corrupted.

For the first test of the day, I went to a small room, put my chin in a cup, and peered into a machine that tested my visual fields. Once I was positioned, I was given a little handheld clicker and the lights were turned out. The device flashes pinpricks of light throughout the range of your visual field. When you see a flash, you are to click the clicker. It's soothing at first, to be in another dark quiet place, peering into a miniature planetarium, but it seems the stars are wired to one of those switches they use for intermittent Christmas lights. Before long, you begin to doubt yourself, triggering the switch and then wondering if you just clicked a floater. Your eyes dry out and you don't blink for the fear of missing another peekaboo constellation. But it's nice and quiet in there. When I finished and came back into the light, Dr. Olson put dilating drops in each eye and sent me back to sit in the waiting room.

I gotta say I enjoyed the eye drops. I wrote some lines on a piece of white paper while waiting for them to take effect. At first, I could fight back, bring the line into focus as it began to blur. Then the black line began to break apart into what looked like 3-D layers of black, yellow, and red yarn. I took out my notebook and wrote a sentence in looping cursive and couldn't distinguish between the letters. Everything became fat and prismatic. Outside it was raining, and I could hear the hissing tires of traffic passing by the clinic. All those people on their miscella- neous errands, and it struck me as it has before that the first thing you crave when your health is threatened is the carelessness of mundane days. A white blur of a woman at the front of the waiting room leaned through the doorway and said, "Michael?" and I gathered my things, but by the time I got to the doorway, she was leaving the room with an old man. An old lady crept past me with her walker and when she got to the window the receptionist pitched her voice higher and spoke louder and slower, which indeed the old woman might have appreciated, but I found grating. I fear I will not age well and, given a walker, may use it as a weapon.

The next "Michael?" was for me, and soon a woman was taking pictures of the inside of my eyeball, which, if I may say so, is the aural equivalent of stuffing lit firecrackers in your ears. The light hits you like a physical blow, *wham, wham, wham.*

When I finally made it to the examining room and a figure strode in claiming to be Dr. Olson, my eyes were so blasted I just had to take his word for it. He said the photographs had pretty much reinforced what he saw previously with the opthalmoscope, that the blanched patch at the arterial junction remained but had not worsened. He said the visual field test confirmed the blind spot but revealed nothing else unusual. He says, all in all, my right eye appeared to be a little better than my left, but that's as it always has been. When he called in his partner for a look and a consult, it was apparent they were both baffled. "We both agree that you have had a vascular incident," says Dr. Olson. "Something has occurred to disrupt the blood flow at the point where the white lesion is located. What isn't clear is why or how. The white spot looks like it might have been there for a long time, but the rapid onset of your blind spot doesn't jive with slower infections or an edematous process. We call this sort of indeterminate problem a 'tweener.' " (A few years later, when I had a kidney stone, the ER doc looked at the CT scan and declared that it, too, was a "tweener." Those folks at the American Medical Association are slacking off.) He went on to say the injury was consistent with something acute, the lesion with something more long-term. "Whether the stricture occurred as a result of the artery swelling or growing closed or from a clot or other object becoming lodged isn't clear."

The bottom line? Apparently something had floated down my blood-stream and parked in one of the teensy arteries that feed the light-sensitive back wall of my eyeball. The cells died, and the black wedge was there for good. The doc's diagnosis: *"Branch retinal artery occlusion with visual field defect."* After scheduling me for blood tests and an ultra-sound and echocardiogram to rule out a stroke and other problems, he sent me on my way, and I set about the business of learning to ignore the fact that my vision would be heretofore obscured by a silhouette in the distorted shape of a Nabisco Bugles snack.

◇ ◇ ◇ ◇

Before I set off on book tour, we taped a plastic wall map of the United States to Amy's bedroom wall. She is usually long asleep by the time I get free to call, but every night before bedtime stories, Anneliese helps her track my progress with an erasable marker. At three and a half years old, it's tough to tell what her interpretation of all this might be, but we hope it gives her some sense of why I am away.

In Oakland, I attend a release event for the book *Quirkyalone*, and pick up a free T-shirt for Anneliese. She and I have discussed this Quirkyalone thing before, because we are both taken by author Sasha Cagen's definition of Quirkyalone as *"a person who enjoys being single (but is not opposed to being in a relationship) and generally prefers to be alone rather than in a couple."* I have often foundered when people have accused me of fearing or despising commitment. On the contrary, I can look to any number of couples in my life (my parents first among them) who have the sort of relationship to which I aspire—but these aspirations have never overridden my enjoyment of solitude and inde-pendence. Cagen says Quirkyalone is not anti-love, it is pro-love. It is not anti-dating, it is anti-*compulsory* dating. Anneliese and I have dis-covered that in the months prior to our meeting, we were both nosing around online dating services, but not making a move. We were both looking and even longing, but neither of us could muster a sense of des-peration, blessed as we were with full lives and good friends. Everything gets turned into a sweepstakes. Society is always trying to sell you tick-ets. "You'll be a lonely old man," people have told me. Maybe, I say. Or maybe I'll be delighted with my freedom. I have been through the standard convulsions of love—swept up, swept away, swept under. I have, at times, wept at the thought of separation. Other times I couldn't wait to get away. I have lain in the dark wondering how I will ever be happy again. But in time, I always *was* happy again, each heartbreak reduced to layers of thin veneer, or, in the tougher instances, a carbonaceous little ball to be left alone in the deepest recesses of the gut. The lesson seemed to be, take happiness as it comes, don't try to get it cornered or run it down from behind.

As far as I can see, the main drawback to this Quirkyalone business, is, well, designating yourself "quirky." I am reminded of a passage in the book *Gridlock*, by Ben Elton, in which he writes that people who put a sticker on their computer "amusingly proclaiming" that "You don't have to be mad to work here but it helps" are guaranteed to be "dull, dependable and sane as a pair of corduroy trousers."

I spend most of my West Coast time alone, chauffeuring myself around in a rental car. I was delighted at one point to find myself hammering down Highway 101 on my way to San Mateo when AC/DC's "It's a Long Way to the Top (If You Wanna Rock 'n' Roll)" came on the radio. When the bagpipe solo cut loose, I kicked that Hertz Mazda up another ten miles an hour.

After California, I fly to Seattle. Then Portland, Oregon. Then Denver. I'm not big on flying. My palms get sweaty and my chest gets light with every takeoff and stratospheric bump. I've read the statistics on flying versus driving but my adrenal glands aren't buying it. This morning as I packed at 4:00 A.M. before catching a taxi to the airport, I turned on the television and caught a documentary about the Red Baron. This struck me as inappropriate preflight viewing. Once while waiting in the boarding lounge before flying out on a writing assignment, I opened a magazine to the first page of a comprehensive postmortem piece on the crash of ValuJet Flight 592. I set it aside and boarded the plane only to find the fellow seated ahead of me holding a newspaper open high and wide to a full-page spread documenting the crash of Swissair Flight 111.

Eleven days into the tour, I am in St. Paul. Amy and Anneliese are driving over to visit me. I have several afternoon events, but my evening is free. I am alone in the hotel room when I hear Amy's little knock on the door. It is like Christmas when she runs in and jumps into my arms. Anneliese follows and we all hug. It is an oasis of an evening. We all cluster on the bed and watch *Daddy Day Care*. My publisher sent over some treats including a chocolate-covered strawberry so big Amy has to attack it apple-style. Later, when Amy is asleep on the guest bed, Anneliese and I sit shoulder to shoulder. I give her the Quirkyalone T-shirt and we both look over the book. We are relieved to find there has been a category established for Quirky*togethers*.

In the morning we have time for a quick walk across the Wabasha Street Bridge. Anneliese and I sit on the green grass of Raspberry Island while Amy stands at the center of the Schubert Club Heilmaier Memorial Bandstand and gives three rousing renditions of the ABC's. The bandstand is a clean wooden plane set beneath a glass and steel arc, a perfect frame within which Amy sings, "*. . . next time won't you sing along with meee!*" and then steps forward to curtsy. When the courtesy van pulled away from the hotel twenty minutes later and I saw them hand in hand, walking for their car in the sun, I had a taste of what it is to love someone so deeply you are terrified for them to be walking the face of the dangerous world. When the jet planed out at 35,000 feet and set course for Chicago, we were well over Wisconsin and the sky was so clear I could trace the highways below northward to where a flat collection of squares clustered on the vanishing point of the horizon: New Auburn. Surreal, in the midst of a month like this to look out a plane window and see your hometown. It felt all the farther away for being within sight. By nightfall, I was hailing a taxi in Pennsylvania. Anneliese and I have been together six months now.

○ ○ ○ ○

All things are relative to a point, and then they are a concrete block to the face. My niggling infirmities likely point to nothing more profound than standard progression. Shoot, even if you lose an eye, you've got another eye. My friend Ozzie, on the other hand, is permanently excused from all equanimous waffling. "People ask me, 'How'd you get in a wheelchair?' " he says, "So I give'em the short version. I tell'em I was on my honeymoon and I got in a car accident."

Ozzie is a ventilator-dependent quadriplegic. He cannot breathe on his own, and he has no movement or feeling from his chin down. He can run his electric wheelchair by blowing and sucking on a straw, but he needs a nurse at his side twenty-four hours a day. And since he lives pretty much in the middle of nowhere over in the county next to mine, help can be hard to find. I was nervous when I first started working for him in October of 1997. When we talked on the phone I explained that while I still had my RN license, I hadn't worked in a hospital setting for years. He said I'd be fine. Later, I came to realize that Ozzie's definition of "nurse" was not as hidebound as that held by those uncompromising sticklers down at the Wisconsin Department of Regulation and Licensing. The folks I met coming on and off shifts ran heavy to tattoos and probation. Ozzie called more than once to see if I could cover at the last minute until his scheduled nurse made bail. The parade of friends that came and went covered the spectrum from trucker to biker, and I mean to include the men as well. You haven't really explored the outer limits of health care until you've watched a Hell's Angel suction a tracheostomy tube. I was, as it turned out, undersized and overqualified.

Ozzie was a gearhead. Since the day he could drive, he ran hard. "I always thought of myself as a good driver, but I was definitely a lead-foot and liked the speed." When he joined the army out of high school and shipped out to Germany, he got hooked on the autobahn. "Speed is nothing over there. It's just a normal everyday thing, to drive fast. I wouldn't call it reckless. Just something you became accustomed to, driving back and forth to work. I was always running eighty to ninety miles an hour." Ironic, then, that when he finally wrecked, he was going 30 miles an hour with his seat belt on.

"I met my wife-to-be in Houston on February 18, 1988. We hit it off right away." Ozzie had been shipped home from Germany and was now stationed at Fort Bliss. "In January of 1989, I reenlisted for another two years, and that year went smooth, and we ended up getting married in July. She already had a son."

When Ozzie talks, his sentences are filled with stops and starts. He has to speak to the rhythm of his ventilator, has to pause while his lungs are forced full of air, then say what he can in the time it takes the air to flow back out. People—especially on the phone—are always cutting him off, thinking he has finished or lost his train of thought. He is the victim of serial interruptions. Even when he laughs, he laughs in silence until the air reverses direction. It is as if he is force-fed an ellipsis every five seconds.

"We got married in Texas. Spent the night in the Radisson. The honeymoon suite. The next afternoon I packed up the car, and the next morning we left for Wisconsin." Ozzie made the trip in a day and a half. "Drove straight through, in my old Gran Prix, My mother threw a reception for us, at the Lucky Lady in Cameron. Invited all my friends." He had enough leave so he and his wife could spend a couple weeks in Wisconsin, and while he was nosing around, he came across an old Ford Bronco with a FOR SALE sign on it. "The body was pretty well rusted out. The drive train and the running gear, there was no problem with that. It started right up and went down the road straight." This is what you call a "beater." A disposable vehicle. You run into trouble with a beater, you just get another beater.

"I was towing it to Texas behind the Gran Prix. We took a shortcut on this old country road between highways. The road was very rough. I got in the back, in the Bronco, to keep it going straight down the road, and hit a pothole. The tie rod broke, and the vehicle flipped up on its side, and the roof of the vehicle hit me on the back of the head and broke my neck. I was going about thirty miles an hour. The seat belt broke loose and threw me in the ditch. This feeling of numbness ran through my whole body. And it was a bizarre feeling, like my foot had gone numb, and yet it was my whole body.

"My wife run up to me. She said, 'C'mon Oz, get up, you're all right.' She thought I was playing. I was a kidder, y'know.

"My last words was, 'I can't breathe.' And that was all I could get out."

He says he remembers it was a bright sunny day because he had to squint at the sky, and he remembers lying on his back and hearing a car approach, and hearing the sound of running feet, and then the sunlight narrowed and went black. He remembers waking once to the feeling of strong wind on his face and the noise of a helicopter and then nothing again. Then he remembers being in midair, and a man in a helmet leaning in to say *Hang in there* and then, again, nothing.

"The next time I woke up I was on a table. This nurse was saying, 'What is your name?' And I couldn't say anything. But I kept mouthing the word *Ozzie*. Then I finally tried to tell her *Look at my arm, look at my arm*. And she kept saying, 'No, what is your name!' She didn't get it—my name is tattooed on my arm. Then I heard a drill-like sound in my skull. They put ice-pick traction on my head to stretch out my neck." He blacked out again. "The next time I woke up I was on this turning bed in a private room with about a dozen machines all around me beeping and flashing and blinking."

When his hair is long and his beard is full, Ozzie resembles your standard midwestern Jesus. That is, if Jesus played a lot of Judas Priest on a sound system with speakers the size of twin doghouses. He controls everything in the room—the lights, the television, the stereo, the louvers on his skylight, the telephone—with a straw positioned at his lips. The first time I took care of him, he sent me out to get the mail. His mailbox is at the end of a relatively long driveway, and I asked if it was okay to leave him alone. He grinned and blew a code into his straw. The two speakers at the foot of his bed began to pulse, and in no time you could feel the floor joists trembling. The speakers were at about a quarter volume. "If you see the roof bouncing," he said, "get back in here."

He spent three weeks in the hospital. Early on, his Wisconsin family was told to hustle down because the doctors weren't sure he'd make it. After three weeks, he was transferred to an army hospital. Ozzie says the hospital was small and they didn't seem to know what to do with him. Within three weeks he was transferred to a rehabilitation facility in California. During his admission physical, doctors discovered thirteen open skin ulcers, three requiring surgical repair.

He spent fourteen months in rehab. Says he lost a little over a hundred pounds from a six-foot-two frame that weighed barely two hundred pounds to begin with. His wife and his mother fought over his care. "I asked my mom to go home, to let me be, because I am a married man now," he says. "She was heartbroken. I know they both wanted the best for me. I told her I loved her and I'd see her again soon after I got settled."

In November 1990, he went home to Houston with his wife. She had rented a home, obtained a van with a lift, and assembled all the necessary medical supplies. When Ozzie returned, it was to a warm welcome from his new in-laws. "But that house wasn't wheel-chair accessible at all," he says. "Very tight corners, very hard to maneuver. I did some damage there."

By June he knew it was over. "I tried to make things work with me and my wife. She tried so hard to take care of me. And her mom and dad helped out, but they could only do so much. Every day she would break down crying. She was crying day in and day out the whole time. She did a lot of drinking. But so was I at the time. She would give it to me, or my friends would give it to me when they came by. We had a drinking problem.

"One time I was watching VH1 and this girl came on in a video and I said to my wife that's a neat song and she just lost it. She said, Oh, you just like the girl. I said no I don't care for the girl, I just like the song. To change the subject I asked my wife if she'd scratch my eyebrow. She said why don't you have that girl scratch it for you. Joking, I said, I would, but she's not here in the room with us. She slammed a Bible on my chest and stormed out.

"I wasn't being turned side to side, wasn't getting my bladder drained because at night she wouldn't wake up. The alcohol, I'm sure. She was sleeping on the floor the whole time near my bed. She refused to get her own bed or share a bed with me."

He pauses for a moment. "Her story would probably be different."

$$\circ \; \circ \; \circ \; \circ$$

Sometimes I wonder what Ozzie used to sound like. Just as the length of his sentences are dictated by the pulse of his ventilator, so is the tone of his voice. He has to speak with the air he is given. We had known each other five years before the day he told me the whole story of his injury. We were sitting on his back deck, and the whole time his voice never changed timbre. Not even when one of his nurses—in jail, the last we heard—showed up going 90 miles an hour on meth and went into a mad dancing rage. He stood ten feet from Ozzie, threatening violence and ranting about a chain saw. He had it in his head that Oz had turned him in for his last offense, never mind that every sheriff's deputy who had worked for the county more than twenty minutes probably knew him by first name, last name, and middle initial. He was spewing out paranoid threats so rapid-fire he was actually casting spittle. I moved a little closer to Oz in case the guy lunged, but I won't lie, I was terrified. Just as suddenly, the madman spun on his heel, jumped in his truck, and peeled out of the driveway. I couldn't get over how calm Oz had been. Then I realized he had no way of acting otherwise. Ozzie rarely complains, but he did say once that he counts among his frustrations the fact that when he is angry with someone, he has to talk to them the same as if he were asking for a glass of milk. He longs, sometimes, to yell.

Back in Houston, Ozzie said he wasn't getting bathed properly. He was missing more meals and starting to lose weight again. The drinking was getting worse. He says he would hear friends and once his pastor came to the door, and his wife would tell them he was sick or sleeping and couldn't see them. Friends would call and she wouldn't put him on the phone. When he was first in the rehabilitation unit, someone told him that only one in five hundred people survive an injury as high up the spine as his, and he says his first thought was *Well, that's too bad for the next 499.* But now he believed he had more than a marriage problem. He believed he was going to be neglected to death.

"She let me talk to my mother on the phone once, and when she left the room I managed to tell my mom I had to get out. We set it up where I got a nurse to take care of me. I told my wife to take the boy to church that morning, and my mom and her husband and some friends were waiting around the corner with a van. They hauled me out and put me on a mattress in the back of the van. Everything else of mine they put in a Budget truck. My mom left a note for my mother-in-law.

"Then we headed'er home."

From Pennsylvania the tour has taken me to New York, Vermont, New Hampshire, and a fair swath of Michigan. I have had three hours of sleep and been up since 4 A.M. when I come down an escalator in the Detroit airport and wind up eye to eye with a friend I haven't seen for over a year. He's in from Seattle with another long-lost mutual friend, and we're all three flying to Madison, Wisconsin. On the same flight, as it turns out. The plane is nearly empty, so we cluster up and chatter until we land, then debark and go our separate ways.

A week of touring remains, but it's fun to be in Madison. I'm here for the Wisconsin Book Festival and it feels like a home-turf time-out. Best of all, I'll be here for three days, and as soon as she finishes her last class, Anneliese is making the three-hour drive down so we can sneak part of a weekend. At the hotel, I see another basket of flowers and goodies including a bottle of wine, and I think, wow, my publisher is really laying it on, although the wine is a little strange since they don't know Anneliese is coming and they do know I don't drink. I unpack my things and then, although I am scruffy and travelworn and need a shower and shave, I set off for downtown with the idea that I will get a card and flowers.

Just off State Street, I find the flowers and get a card, then begin the walk back. It's sunny but crisp, and I am wearing a heavy, oversized sweater over my T-shirt. I am scuffing along in my steel-toed boots and approaching a residential intersection when I spot a large Ford Explorer approaching from the left. A college boy in a backward baseball cap is at the wheel. He is coming from a feeder street and has a stop sign, so I am half into the crosswalk when I see him looking at me, then up the street, then back at me, then up the street the other way, and I realize he is doing math. Calculating speeds and vectors. He's going to shoot the stop sign, I'm thinking, when *Brrrrmmmm!* that's exactly what he does. We've made eye contact twice, so this isn't an issue of him not seeing me. The front wheels just miss my boots, and he is so close I can feel the push of the air moving off the quarter panels. His venal disregard for my safety in the interest of divesting himself of five seconds at the stop sign triggers something in me. The rear quarter panel of the Explorer is right there, passing a foot in front of me, so I just haul off and kick it as hard as I can.

It makes a delightful *whomp*.

He does a brake stand, stops dead smack across the centerline, and whips his head around. I am panicking at what I have gone and done, but he doesn't know that, so I give him my best deranged hillbilly glare from beneath my ominous unibrow. I think it helps that I haven't shaved, but in the end I think what really sends him on his way is me clutching that big bouquet of flowers. He snaps his head around to the front, tromps the gas so hard pure springwater flushes out the tailpipe, and speeds off to whatever keg party it is he's late for.

I pause for a cleansing breath and walk on. I am ashamed and elated.

Anneliese and I sneak two wonderful days, going to readings, dining with writers and book folks, and walking hand in hand up and down State Street. When I got back to the hotel after kicking the car, I had time to clean up before Anneliese arrived, and it felt grand to open the door and welcome her into the room. After we sat and talked and caught up for a while, I asked her if she wanted anything from the goodie basket, in particular, the wine. She looked at me quizzically. "Who is it from?"

"My publisher, I suppose," I said.

"Did you look at the card?" she asked.

"Um, no."

"Well you might want to look at the card."

I opened the miniature envelope and pulled out the miniature card. It was from Anneliese. To celebrate six months, she had written. She also mentioned the future. When I looked up, she was smiling, but she had one eyebrow cocked. "Check the wine," she said. It was nonalcoholic.

We had the wine then, and talked about that future, and how at six months it seems like six years in a good way, and then we talked about maybe one day having chickens. It always comes back to chickens. Then she drove home and I flew to Kansas City.

Out on his deck there, overlooking the old farmstead where he now lives, Ozzie said the doctors at the little army hospital told his wife he had two good years to live. The doctors at the rehabilitation facility said ten. He's made it to sixteen. "Three-quarters of the guys I went through rehab with are dead now," he says. "And they had lower-level injuries. They could use their arms and hands and stuff, which I guess when you look back on it might be worse, because that's how they took their sleeping pills. To finish themselves."

Ozzie and I have had this talk: I've told him sometimes I think of myself in that bed, crazy for someone to itch my eyebrow, and I tell him I'm not sure I wouldn't ask that same person to give me the sleeping pills. I imagine what he feels to be akin to all-consuming claustrophobia, and I think I might be too weak to take it. More than once, after I've finished suctioning an obstruction from his trachea (a process he endures many, many times a day, and which is absolutely essential to keep him alive but puts him through choking fits of coughing every time), I will be heading for the trash with the gloves and used tubing when I will hear a soft clicking noise. It's Ozzie, using his tongue to make a soft *tsk-tsk-tsk* against the back of his teeth. He teaches you this one your first day on the job. It's his signal that you've forgotten to reattach his ventilator tube. He can't scream, he can't grab you by the neck, he can't stick a boot in your terminally forgetful ass the way he ought to, he just has to patiently *click*. "I don't have any fear of dying anymore," he says. "I've had a couple close calls where my tubing's come apart and I've had to wait at the most maybe two minutes. Just because a nurse might've gone to the mailbox. I just kind of accept that when my time comes, it's not within my control.

"Sometimes people ask what things I miss the most. If I had to give my top three things, I'd say, first one is breathing, of course. Second is, I'd love to be able to wipe my own ass. Sounds funny, but it's true. It's a personal thing. And my third thing, is drive motor cycle again. That was one of my greatest loves, to get on the road on a bike."

Ozzie has a friend who repairs motorcycles in a little shop west of here. Pat is a paraplegic. He has a bike set up with hand controls. He hoists himself on and off the cycle with his arms, and stows his wheelchair on a metal grate attached to the cycle like a sidecar. He visits Ozzie pretty regular and he knows how much Ozzie would love to ride again. "I should take you for a ride," said Pat one day. He pulled the motorcycle into the ditch so Ozzie could drive down onto the grate. His electric wheelchair weighs over four hundred and fifty pounds, so Pat wedged the wheels with a couple of two-by-fours and belted the whole works down.

"I asked him to start off slow," says Ozzie. He admits he was nervous. All that open air, nothing around him. "I'm going down the road," he says, "thinking, 'What the hell?!?'"

But he says it was so good to feel the wind in his face.

They made a three-mile loop, flying down the road, two bikers riding free, one good pair of arms between them.

Ozzie believes—hopes—he is in this predicament for a reason. He says this is what keeps him going, and he is at peace with whatever will be. But one unanswered question haunts him, and has since the day his Bronco pitched him into the ditch. He knows the running feet he heard just before the sun went out belonged to an off-duty EMT who just happened to be following him down that Missouri back road. It was the EMT who kept him alive, breathing for him until the helicopter got there and the flight nurse could bag him with an endotracheal tube. He knows it was an EMT, because his mother got the story from someone at the hospital who said an EMT stopped in to see if Ozzie was alive. But the person never left a name. He doesn't know if it was a man or a woman. "There was no name in the police report," says Ozzie. "I'd sure like to thank that person. I even went so far as to take an ad out in the paper down there. That was ten years ago. No one responded. My mom said maybe they were afraid of getting sued."

◦ ◦ ◦ ◦

The last time I saw Dr. Olson, he said all my tests had come back normal. He dilated my eyes again, had a look, compared what he saw to the pictures taken previously. He said the lesion was reducing, which suggested that the episode was indeed acute. None of this resolves the mystery of what happened, if there was a clot, and if so, where it came from and why. Fortunately, the eye is an organ capable of deception. In collusion with the brain, it convinces you to ignore what you see—or don't see. Over time, the blind spot has faded. It no longer bothers me when I read. In fact, I have to be in the perfect setting—standing in a wide field of snow, say, or inside a white shower stall—in order to see it, and even then I have to gaze up and to the left and concentrate intently, or not at all—the way you do with those hidden 3-D picture books—before it reappears, a hazy gray wedge. It could be worse. Shortly after I found out the blind spot was permanent, I was explaining the situation to my brother Jed.

"Which eye?" he asked.

"My left."

"Well, at least it ain't your shootin' eye."

Which is true. My sister-in-law Barbara had the same sort of event, only her blind spot is in the dead center of her right eye. When she looks through a rifle scope, the critical spot where the crosshairs meet disappears. She has had to learn to shoot right-handed using her left eye, which is like trying to drive your car mailman-style. It takes awhile before the reprogramming kicks in.

Every once in a while when I am in a dark room, or drifting off to sleep, I see a little flash in my left eye, as if a teensy paparazzo is firing in the distance. It's a warning light, a reminder that the game can expire at any time. When I recall the period immediately after the blind spot arrived, what comes back most clearly was my desire to just get out of the waiting room and back to everyday chores. While he was waiting for the return of definitive test results, Dr. Olson restricted my activities, and I remember sitting in the fire hall when the other members were roaring off on a call and how badly I wanted to sling on the gear and go, just to run and sweat and bull against the hose.

When I first started working for Ozzie, I found it strange to hear him say things like "I cleaned the garage; I rearranged the kitchen; I keep those boxes in the basement; I changed the oil in the van." It didn't take me long to realize that his use of "I" was deeply intentional. The moment you come on shift, you become Ozzie. The tasks you perform are Ozzie's tasks. I have caught myself—when he interrupts my reading to summon me from the other room to adjust his headset, or feed him a boiled egg—feeling a small bristle of irritation at his voice in that split second it takes to reorient myself to the fact that for twenty-four hours my legs and arms are not mine. Ozzie's use of the pronoun "I" is a clear declaration of independence.

I used to take care of Ozzie at least once a month. This was not a charity gig. I got paid the same as all the other nurses. For the past year, I've been on the road so steady I haven't been able to take any shifts. I do try to call him now and then, or he'll call me, and I like to send him a postcard, especially now when I'm out on book tour. When I'm home, I try to keep Ozzie posted on how things are progressing with the resurrection of the International. He likes to hear about that stuff, and I bring him pictures now and then. I've promised him a ride when I get it running. We'll set up a ramp and run his chair right up in there. Can't be any more dangerous than that sidecar.

Ozzie's got his own project going, a '68 Dodge Charger he's restoring. He had one in high school. This one he's making into a drag racing car. "This guy I'm working with, he's had my car for ten years. He keeps whittling away at it whenever I can afford it. It's got a 426 engine, electronically fuel-injected and blown, it's got wheel tubs in it, a six-point roll cage. I guess when it's all done, I'll hit the car show circuit and show it off, show how my dream came to realization.

"I remember having a conversation with my mother, and what would happen if I died, and I jokingly told her if I do, I would like to be cremated and have my ashes dumped into the fuel tank, and go down the drag strip one last time, for my last ride. Of course, I was joking, trying to make light of the situation. Mom kinda cracked a smile. Shook her head, like, *Okay*."

· ○ ○ ○ ○

I declare that my blind spot admits me to the One-Eyed Club, albeit on a technicality. I recognize that the step from losing part of one eye to losing the entire eye is a biggie (which in turn does not compare to the step from partially blind to blind). At best, mine is a provisional membership.

But I'll take it. There is this ridiculous little part of me—residual of the ten-year-old who posed lurid drawn-out death scenes beneath the bird feeder—that likes the idea of being the Hathaway Shirt man, slightly dangerous and mysterious behind the black patch, soldiering nonetheless dapperly on. When you look into my eyes, not all of me is looking back. I am a stoic man of mystery. People sometimes miss this.

I'm in a hotel room overlooking St. Louis. It's late, and I can see the Arch, all lit up and steely. Tomorrow I fly home. I am missing Amy and Anneliese. Every time I take off in another airplane, right when we are reaching top speed on the runway, I imagine them in my arms, Anneliese on my left, Amy on my right, both with their head resting on my shoulder, and I can make it so real I can feel the curve of their ribs and the warmth of their skin and the scent of their hair. Ozzie told me recently that his wife still calls him sometimes. They were divorced years ago, and she has remarried and divorced twice since. But she still calls. Always after midnight, he says. I'm sorry for the way it had to be, he says, but I know if I had stayed I would have been dead a long time ago. She would probably not agree. But I think that would be the case. I visited her parents last year, and her son. I told her she should come visit, but she chose not to. So I never did see her. I was eighteen days into my marriage when I had my accident. And the weird part was, me and my wife had only spent one night in our house together.

Chapter 12

NOVEMBER

A LONG ABOUT THE second week of November, the men of Wisconsin begin to go scruffy. You'll notice it everywhere—at church, at the gas station, in the Wal Mart—even the jawline of the local banker begins to blur. I am no different. We are the men of Wisconsin, and we are growing our deer hunting beards. The deer hunting beard protects your chin from the chill air and staves off windburn. The deer hunting beard preserves the brotherhood and scratches our women. The deer hunting beard reminds us why most men should keep at it with the razor.

I was years from my first beard when I first skipped school for deer hunting. Grandpa shot a buck out on the west forty, which in those deer-poor days was a remarkable occurrence, so Dad let me run out there to see it. By the time we got it gutted, hauled to the house, and hung, the school bus had come and gone. I remember when I walked back into Mrs. Kramschuster's third-grade classroom I felt chesty and important, as if I were returning from a manly mission. It is no wonder young men go so easily to war. By the time I was old enough to buy a hunting license, the school board had given up and just shut the place down for the entire nine-day season.

Mark and I hoped to have the truck ready by deer season. It's not going to happen. We're getting close. Or more to the point, Mark is getting close. While I was on book tour all last month, Mark reattached the bed, rewired the six-volt system to twelve, and painted the box and body. When Anneliese, Amy, and I went to Mark's house for our annual family Halloween party last night I saw the truck sitting there under the yard light and was flabbergasted at how good it looked. Mark used a flat marine green paint, which is perfect because it doesn't reflect the light, and all the dings and wrinkles recede. He's done the brush buster, the bumpers, and all the trim in black. The combination is pleasant to look at. Clean and calm, nothing flashy. Under the mercury-vapor light it looked like the truck had rolled right off the set of *M°A°S°H*. We stood there and admired it for a while, me a cowboy with a fake mustache, him a vampire.

◇ ◇ ◇ ◇

On the third of November, we receive our first snowfall. I am running errands with Amy belted in her car seat and to my everlasting shame I am sneaking a listen to the local sports talk radio station. The NFL show. The host is assessing the play of the Seattle Seahawks and from the back of the seat I hear Amy's happy little voice: "That's where you were!"

Wow. Yes. Four weeks ago. She was asleep when I called from my hotel room.

I think about the map on her bedroom wall, the one with all the lines and circles.

I turn off the blankety-blank radio.

I get over to Mark's place once to help outfit the International with a new radiator and heater hoses and refill the antifreeze. We've been frustrated in our search for a pair of windshield wipers, but while we're talking in the parts store it hits me that there may be a pair on the L-180. Thinking Kathleen might have tired of looking at the fenderless hulk in her driveway every morning when she came home from work, I had my brother Jed bring it home on his equipment trailer one night when he was working with Mark, and now it's stored behind Jed's barn. I'll swing out there during deer hunting.

The running boards on the old truck were beat up and bent, and rather than try to straighten, repair, or replace them, Mark has suggested I opt for nerf bars—essentially a doorstep made from tubular steel and mounted to the frame rails beneath the cab. You can order them premade, but they're pretty spendy. Mark says he can make a pair if I just buy the raw iron stock. We decide to go this route, although it means the cool silver exhaust pipe will now extend out into the middle of nowhere, and will have to be cut back and replaced with something more modest.

But the thing that has us grinning right now is that the new seats are in. They are figure-hugging stock car seats. We bought them from a mail-order company that sells racing equipment. We ordered the ones with four-point restraining belts. There was actually some reasoning here—the seats are thin and create a little more legroom for Today's Man inside the relatively teensy Comfo-Vision cab, and the belts are a genuine safety upgrade, but mostly we got them because they look so delightfully silly in that fat bug of a cab, in a truck geared to lug fourth at 10 miles per hour. They make it difficult to get in and out, but someone once said trucks should not be easy.

$$\circ \; \circ \; \circ \; \circ$$

People ask you sometimes when you knew you were falling in love, and I have the answer. Back when we first met, maybe the third or fourth date, Anneliese picked me up in her car, the battered little Honda. When I ducked my head and dropped into the worn seat, she was apologetic. "I know I should probably drive something that's in a little better shape," she said, "but I'm too cheap." I don't remember what does it for you when you're twenty-five, but when you're self-employed and crowding forty, that kind of talk makes you want to skip around and fling daisies.

She's not cheap, of course. She's frugal. But I liked her even better for choosing the word *cheap,* as it spoke to a certain kind of raising. I drive a used Chevy not out of some self-depriving morality but because the money arrives in fits and starts and I want to minimize the amount of time spent working for the financing company. When it comes to love, I am told cash flow is a leading cause of cancer, and in this Anneliese and I agree: low overhead is the key to survival. You will notice I fail to itemize the costs related to the repair of the International, as this would be unpoetic. Montaigne has earlier noted our capacity for contradiction.

I'm still wrapping up the book tour, and over the weekend I did a little spate of events in Kentucky and Illinois. At one point I was hosted at a wine and cheese party attended by a woman of some academic tenure. While I was cruising for cheddar in the kitchen, I could hear her in the drawing room, comparing the attorney general to a pair of Nazis. She was particularly dramatic on the issue of privacy, becoming visibly exercised regarding her confidence that right now someone was preparing to peek in her bedroom and plotting to pull her library card. "Terrifying!" she kept saying. "Terrifying to have these people in power!" Fair enough, and pass the brie. But my interest was piqued when five minutes later she declared she didn't understand why—if people *had* to own guns—why they would be loathe to submit that information in written form and accept some "reasonable government oversight."

Suddenly Himmler and Goebbels are Andy and Barney.

I held my peace, as I am a polite guest and a coward and hadn't had any of the wine, and cheese makes you peaceful, but perhaps due to the impending deer hunting season her paradoxical take got in my head and I chewed on it for a while. I first squeezed the trigger of a gun sometime around the age of nine or ten, doing so under the direction of the same father who forbade us to own toy guns or pretend to shoot each other with our fingers. Naturally when we visited friends we dove straight for the plastic pistols and went full-on OK Corral. That is, when our friends could pry us away from their television (also banned from our house and to which we were consequently drawn like lost Amazonian tribesmen to a functioning Lava Lamp). Before I ever touched a rifle Dad repeated two simple rules, over and over: Treat *every* gun as if it is loaded, and *never* point it at another human being. Then he showed me how the gun worked, how it came apart, how to check the chamber with your finger and your eyeball to make sure it was empty, and then once you had assured yourself that it was empty beyond a doubt, to treat it as if you were certain it was loaded and ready to fire. He taught me never to switch the safety off unless you intend to shoot, and never shoot unless you have identified your target. He taught me how to line up the front sight and the rear sight, and he had me watch while he fired. When we looked at the punctured steel cans and the chunks blasted from the log, he made sure I understood the destructive power of a piece of lead half the size of a pencil eraser. Only then did he hand me a live round. I chambered it and fired it under his watch, and understood I had been given responsibility for a potentially deadly tool.

We took Dad's admonitions seriously and I can't recall a single instance of horseplay, although we did shoot grackles out of the tops of pine trees, and if we missed, the bullets would go whining through the sky to who knows where. Sometimes when I went to bed I'd lie awake thinking maybe a round had dropped from the sky and winged one of the Teed family. If by morning there was no news, I resumed life with a clear conscience.

I own three rifles, two shotguns, and one revolver, which is probably low average for my geographic peer group, and leaves me in Ted Nugent's dust along with all the rest of you, but puts me well ahead of other outspoken celebrities who believe guns are evil unless they are rented with bodyguard attached. One of the rifles is a .22 semiautomatic; the other two are 30–06 bolt actions. Both shotguns are twelve gauge; a pump action and a single shot. The revolver is a rather gigantic Ruger Super Redhawk .44 Magnum with a seven-inch barrel. I got it after several recent bear encounters. In the old days, you saw a bear, you yelled at it, it ran off. These days, with more houses and more people and more garbage cans and feeding stations, they are not so much frightened anymore. Recently my brother John was unable to open his front door, as it was blocked by a bear. When he pushed the door open, the bear stuck his nose in the crack and tried to come in, which you could say forced the issue. I am not panicked by bears, or I wouldn't be walking around the woods. Ninety-nine times out of a hundred, a bear in the wild is going to run the other way. I have no desire to shoot one. But neither am I willing to test the *just play dead* theory with Bear Number 100, especially if he comes disguised as Bear Number 17. I don't recommend handguns in general, as they handle like sports cars and similarly tempt you to operate a little faster than you should. And buying one for home security is silly on the order of sweeping the sidewalk with a feather duster, although with a feather duster you might actually hit something.

But if you're working in the woods, it's easier to strap on a sidearm than tote long iron.

It's hard to talk about guns without sounding defensive or blustery. I'm pro-gun the same way I'm pro–potato fork. I use them both to gather food for the year, with the caveat that if you break into my house, I won't be waiting for you at the top of the stairs with a potato fork. And even that last comment I offer knowing that I'm a heavy sleeper and will probably get into action way too late, because although the shotgun leans against the wall beside my mattress, it is unloaded and trigger-locked, with the key and shells stored in two separate locations within the bedroom. This quite intentionally impedes the likelihood of a drowsy quick-draw. But all the well-reasoned arguments against providing your own armed defense tend to go a little pale the first time you stand in your own dark house watching some guy get stomped—not beaten up, stomped—outside the bar in the middle of Main Street and thirty minutes pass before the cops show because they are geographically overstretched. Having time after time seen the results of violence—including deadly gun violence—so close I could smell it in the back of an ambulance, I go out of my way to live like a peaceable fraidy-cat. But when it comes down to my front porch, I tend to vote with Teddy Roosevelt. Here in Wisconsin there's been a strong effort to establish a law giving properly permitted citizens the right to carry a concealed weapon. The only thing I find less convincing than the arguments for a law like this are the arguments against it, and if it ever passes, I'll apply for a permit only because I think a guy is silly not to avail himself of all options. At the moment the point is moot, as the law has been vetoed, and furthermore, while it is possible to carry a Ruger Super Redhawk .44 Magnum, concealing it is out of the question.

Perhaps the potato fork allegory will not hold up under scrutiny. Perhaps a better way to put it is that there are legions of us out here who have guns and have always had guns, and we attach to this all the dramatic significance of having silverware. Once when I was standing beside my brother John at his sawmill, our fire department pagers went off and called us in to stand by with the county SWAT team. "We have a report of a man holed up in his house with a gun," said the dispatcher. John looked at me quizzically. "*Hmmm* . . . ," he said. "That's me every night!"

Anneliese is in the swing of the university semester, and I am deep in magazine deadlines. We e-mail each other more than we see each other. We are working around the edges of how we might combine our lives. I have told my friend Gene I am certain this is it, but then when Anneliese invokes the phrase "formalizing the relationship," I'm not sure how to respond. I'm not put off, or short of air, I'm just not sure what the next move should be.

She is game but not overly enthused about my commitment to the ten-day Wisconsin Gun Deer Season as administrated by the Wisconsin DNR (Department of Natural Resources, although some locals will tell you Damn Near Russia). I readily acknowledge her reservations but am pretty much inflexible on the issue. For a rampant skitter-brain like myself, deer-hunting season is my one consistent source of reorientation. Since that day in third grade, I have not missed a season—no matter where I was living at the time or what kind of job I held. In our family this is a tradition handed down over at least five generations that I know of, and life will not unravel if one year I have to be somewhere else, but that week of trees and swamp adjusts my head and puts food in my freezer, and for now I am not prepared to miss it.

Let us not, however, fool ourselves into yodeling golden ballads about carrying on the primal traditions of the hearty provisioners of yore. A lot of hunters around here set up cameras along the deer trails and attach them to a motion sensor so they can learn which deer are working where. If you run into these guys down at the gas station they'll pull the photos off the dashboard and show you. You can pick out the deer immediately—they're the ones with glowing white eyeballs. A neighbor set up one of these devices during bear season and when he picked the film up at Wal-Mart he had several excellent pictures of himself filling the bait station. You can buy digital shooting scopes and range finders, GPS units have replaced the ol' bubble compass, walkie-talkies have replaced frantic waving, and at the extreme edge of things laws are currently being established to address the idea of remote hunts via the Web, in which your weapon is a wireless mouse. My rifle barrel is made of stainless steel and rests in a synthetic stock. I use a scope. I hook a slab of flexible foam to my belt so I always have somewhere soft to plant my hinder.

The real changes are less about gear than land, which is being divided into smaller and smaller fractions and is less and less available for hunting. More and more hunting is being organized into a group or preserve-based pay-as-you-go situation. Some of this is driven by the desire for trophies, which permeates hunting top to bottom and leads to the deer herd being cultivated and manipulated to grow more and bigger antlers. I like a big rack as much as the next guy, but readily shoot any small buck that presents itself, and am thus razzed at fire department meetings by my friends who practice Quality Deer Management, whom the last time we discussed this generally agreed any set of horns wide enough to accommodate a case of beer is a keeper.

Those who cast aspersions sometimes portray the average hunter as a beer-swilling lout incapable of ambulating more than five feet without the aid of a four-wheel-drive ATV and a packet of Slim Jims, to which I say, I dare you to go down to the bar and say that. Of course you will find fellows in blaze orange who have confused their rifles with their penises (and what fun to poll their wives and lovers for the purposes of drafting a list titled Top Ten Reasons the Metaphor Is Inapt, but let's stay on track). You shall know them by their bumper stickers, which run along the line of HAPPINESS IS A WARM GUTPILE or I LIKE LIPSTICK ON MY DIPSTICK. While graceless, neither sentiment is ipso facto impeachable. I find it helps to remember sometimes the First Amendment has a beer belly.

It is easy here to veer off into a defense of hunting and simultaneously bemoan its current state, with its focus on trophies and management and gear and soft, lazy boys, but that is not so much an issue of hunting as it is the ongoing issue of the commodification and overimprovement of simply everything. What I am doing this November is a little bit all of the above. But it is also the most consistent form of local food-gathering in which I have ever participated. I have gardened off and on, I have spent the standard amounts to have animals killed and packaged by proxy, but for the past twenty-six Novembers I have gone into the same patch of woods with a gun and tried to kill a deer so I could eat it.

○ ○ ○ ○

I have taken up a gun with an eye toward my own security only once. Wakened at 3 A.M. by the sound of someone scuffling with the side door to my garage, I grabbed the shotgun, rolled off the mattress, and was shortly at the screen window that overlooks the space between the house and the garage. I got there just in time to see two skulkers enter and pull the door shut behind them. It was a still night. Even with the door closed I could hear their feet shuffling, and I heard one of them say, "I know there's gas cans in here."

They were in for a tough go. At the time my garage was a combination storage locker, dumpster, and recycling repository. Navigation was difficult in the most favorable circumstances

"I know it's in here," I heard one of them say. "I seen him carry it in." Then I heard a faint scratching, followed by two weakly bobbing lights through the garage window. It took a second, then it registered: I was being robbed by thieves using cigarette lighters to locate cans of gasoline.

You'd love to see how this would play out, but with respect to my insurance agent and the rest of the fire department, who need their sleep, I decided to put a stop to it. In my best fake manly bellow I hollered, "You boys better HIT THE ROAD!"

The garage went dark. Then I heard muttering.

"Whuwazzat?!?"

"I dunno."

"I thought somebody yelled."

"I think it was a security system." Now there's a concept. A slantways garage filled with junk, protected by a security system that heckles you.

"No, I thought I heard somebody yell."

There was a stretch of quiet. Then out of the darkness, the same voice, directed against me: "FUUUCK YOU!"

You'll hear that around here after bar time. But what followed knocked me sideways. I heard the other guy speak, quietly but urgently.

"Hey! What the hell! You can't talk to people that way!"

"What?"

"You can't just yell at people like that!"

"Whuhh . . ."

"He might of let us have some, but you can't just holler at people like that."

By this time I had realized they had no idea where I was, so I just stood there silent. Shortly they emerged and made the slow journey out my driveway, up the sidewalk, and started across Main Street over by the Sunshine Café. Slow, because every four feet the thief whose sensitivities had been offended would get in front of his partner, block his progress, and lecture him on civil discourse. They'd been drinking, sure, but the sincerity was touching. The last I saw them, they were in the middle of Main Street in the ambient yellow of the streetlight, and the impolitic thief was saying, "I *know,* man. I'm *sorry,* man."

I leaned the shotgun against the wall and went back to bed. I had completely forgotten to remove the trigger lock.

○ ○ ○ ○

The night before the season opener, I am trying to beat a magazine deadline and wind up writing until midnight. When the alarm rings at 5:30 A.M. this morning, my head hurts, but I drag myself out of bed. Lately it's been snowless but cold, so I layer on my thermals and go downstairs to put the kettle on. I have begun a tradition of taking a thermos of green ginger tea to the stand, if only because I like its fragrance against a clean palate of cold air. This tradition was leaked to the fire department by, I think, one of my brothers and I have suffered because of it.

I quit trying to remember everything years ago and have written up a little list, which I store right in the box with my hunting gear, so everything is laid out and ready to go. Shells, buck scent, knife, paper toweling, notepad and pens, binoculars, hand warmers (you expected hardiness from a guy who drinks ginger tea?), fire department pager and radio (deer season is not the best time for little Susie to light the drapes), trail mix, granola bar, miniature flashlight. When the water boils and the tea is made, I finish dressing: blaze orange bibs, jacket, and hat. Big boots. My rifle is in its case by the door, and I grab it as I leave.

It's a six-mile drive to my father's farm, and all along the way I see kitchen lights on and headlights moving across fields. I drive to my spot and set out from the car. I have to hike across most of a wooded forty, but I am on a logging trail and I walk with the flashlight wedged in my hatband. Among other things, I want to make it clear to anyone who might be out here that I am a human.

The stand is shortly in view, gray in the shifting beam of the flashlight. It's basically a wooden box on stilts. No roof or heater, but walls to keep the wind out, cut to a length where I can sit in a collapsible camping chair and see over. I came out yesterday afternoon and planted the chair and a sleeping bag. I climb the ladder, settle in the chair, arrange my things, and drape the sleeping bag over my lap and legs. When I turn off the flashlight, it is still dark. I can hear the muffled rise and fall of motors in the distance, and the flatulence of ATVs, and now and then through a thin spot in the trees, a set of headlights rounding a corner and moving up a dirt road. Beyond that, there isn't much, and this time of morning the wind rarely stirs. Behind me, to the east, the sky is easing up some gray. The birch and popple trunks are just beginning to emerge as white strips against the black. Things are slowly taking on dimension. I shrug a little deeper into the chair and pull the sleeping bag tighter around my waist.

When I wake up it is overcast but we are in the full light of day, definitely well past sunrise. I check my cell phone clock. 7:30 A.M. I've been asleep over an hour. So much for the wily stalker. I unscrew the thermos cap and tip it full of tea. The wind has yet to move, and the steam rises straight to my nose, fragrant with the ginger. I wait a moment for the tea to cool, take a sip, and then set the cap carefully on the wooden floor. When I bring my head back up, a small buck has appeared on the logging trail standing broadside. He is looking directly at me. If I move he'll bolt, so I freeze. He watches me for a while, then turns his head and looks the other direction, and I raise my rifle and fire. He runs toward me but I can see he is shot through the chest and I do not shoot again. He stops, falls, heaves a breath, and is dead.

○ ○ ○ ○

Back in 1988, when I was first fiddling around with the idea of writing for magazines, I sent a query letter to several hunting magazines proposing that I write an article sharing my tips for a successful crow hunt. I based my knowledge on a sum total of maybe three crow-hunting expeditions with my friend Max Jabowski. His father taught him how to call a crow, and Max taught me. I can still do it. Using nothing but my hand and mouth, I can hide in a stand of popples and have the air swirling with big black birds in under five minutes. I still call crows sometimes when I am out and about in the spring, but I stopped killing them in the early 1990s when I realized it was shamefully pointless and possibly sacrilegious. But what really creeps me out about the rough draft of that proposed article in 1988 (I keep *everything* and am eminently unelectable) is this line: ". . . two hours later we had bagged several crows, with no plans to stop."

Bagged. What a puny little swagger.

I read that line today and am stricken with the usual chronological schizophrenia. Who was this twerp? In scanning the text (Title: "Run'n'Gun for Crows"—pardon me while I seal this airsickness bag), I see we were bombing around in the International, so certain threads remain. But when I look at that little deer lying there now, *bagged* is not the term that comes to mind. As long as you hunt and eat red meat, you will never fully evolve, but I have gotten beyond the insulting claptrap of *bagged*.

Killed. Period. Because I want to eat it. Because this is where the tender stuff in the pan—the inch-thick chop banked with garlic, brushed with soy, garnished with fresh-diced chives, and done just to pink— comes from. Neither poetics nor bravado serves to respect the deer that is done running.

After a decent interval—again, based not on some mystical tradition, but on the fact that where there is one buck there are often others, and furthermore I have a doe permit yet to fill—I climb down from the stand to tag and gut the deer. The registration tag is attached to a larger numbered tag we are required to post on our backs like a license plate for humans. The registration tag tears free along a perforation, and then you have to note the date and time of kill by piercing the proper numbers with the tip of your hunting knife, an exercise that I assume is a leading cause of deer hunter injuries. After further perforating the tag to indicate sex and size, I tie the tag to one of the buck's antlers, then flip him to his back and open his belly. Everyone has their own process here, but the bottom line is, when you are up to your elbows in deer guts, you are shopping for groceries without equivocation. You also cannot help but notice the similarities of the deer's organs to your own, nor is it possible to pass the back of your hand over the jagged edges of a shattered rib without clearly understanding the power of your tools and the violence of your act. It isn't that we are ignorant of what we are doing out here. Sometimes when I hear parlor talk against hunting and guns it isn't the reasoning I dispute as much as the tone. Sometimes I get the feeling someone is trying to give the savages a little religion.

And now I am back in the stand, where I will remain all day with the dead buck at the edge of my vision. It is a good thing in any circumstance, I think, to sit in one place for the duration of the time it takes the sun to cross the sky. You begin to attune yourself. The early stillness is my favorite. Invariably, though, the air begins to shift, and then the noise begins to build. Just now I hear the sound of a squirrel's claw on bark. Then the spiking call of a woodpecker, followed by the soft rappety-tap of pecking. Even with the constant hiss of tinnitus, my ears still work pretty well, and I am listening for the crack of a twig or papery leaf crumple that will tell me a deer is approaching. The distinctions are something you pick up over time. A twig broken high up—by a squirrel, or the wind—has a resonance, whereas the crack of a twig broken against the ground by the tread of a hoof is slightly dampened. A deer approaching through oak leaves tends to generate a steady *swish, swish, swish*. A squirrel working in the same leaves can fool you by hopping to the same rhythm, but when they are foraging, they create a fluffier sound, more like a finger flicking around a loose pile of leaves. Think of the sound of someone looking for the last raisin in the bran flakes.

Visually, these snowless gray days are the least desirable for spotting deer. Everything is tan, brown, gray, or off-white. The young maples have tones of pale lichen green, but beyond that and the needles of the Norway and white pines, that's pretty much it for color. Once on a day like this I did spot a weasel, all white for winter but betrayed in his ermine by all the brown around him. You have never seen a leaner machine. He did not scamper, he poured himself across the ground and over logs like mercury. For ten minutes, he appeared, disappeared, and reappeared. When I finally got the binoculars on him, he had a mouse in his mouth, which made me feel a little better about my hunting.

Some days the sun is bright, which can help or hurt depending on its angle and the cover it strikes. Much of the underbrush around my stand is clad in slick bark, and when the sun hits it from behind you it sets up a myriad of glistening cobwebs, and where you could see fifty yards in the overcast, you can't see twenty. But I once shot a buck that came through that same underbrush in sun of exactly that sort because the light glanced off his horns in a way that didn't match the other light. That's what you're looking for on any sort of day. A line or a movement or a shape that doesn't fit the pattern. Something going lateral when everything else is still or bobbing. Of course when you have the backdrop of snow, you can spot things at ten times the distance.

You will notice rhythms. Often the squirrels come in waves, then abate. Wild turkeys come streaming through afoot, spot you, and go clattering into flight, breaking branches and then leaving everything suddenly silent, although you may hear a gobble from where they have landed. I once watched a partridge feeding for half an hour until he was directly beneath me, at which point, because I do love fresh partridge, I tried to hit him with my thermos but missed wide right, and when the sound of his wings faded I immediately missed his company.

I stay in the stand all day and don't see another deer. I admit I snoozed a time or two, so there may have been a jailbreak. But now I'm watching the day draw down. Twilight in November comes early, pressing at you from early afternoon onward. The cold becomes foreboding, obvious as it is that it will only get worse as the so-far-ineffective sun fades away altogether. Whenever I read paeans to nature—my own included—I am grateful for these late November days, which mercilessly reduce you to the proper size. This is not the lover's sunset of summer with its manic colors and promise of a tantalizing night. The deer hunting sun sinks as if it will never rise again. The light, already anemic, simply drains away. Even if the sky colors up a brassy pink, it remains metallic and you view it through bony, naked branches. There are times right at the edge of darkness when a crow will drift by silently and swerve to disappear into the inky silhouette of a tall pine and you are certain your life is ended. By the time you are out of the woods it is fully night, and back in the underbrush the fat on the gutpile has gone solid.

It is a relief, then, to roll up to Jed's shop and step through the door into the woodstove warmth, everyone gathered after having made their way back out of the trees. We skin and quarter the deer, hanging them one by one from a steel beam and ratcheting the skin away with an old boat winch, cutting the body into five pieces to be hung from hooks. We catch up while we work, everyone relating the story of their day. Later in the week we will begin carving the deer into the roasts, the chops, the parts for sausage, but tonight we just skin and quarter and then go home to our houses where the lights seem bright and everything unnaturally square.

○ ○ ○ ○

We hunt most of the week, sometimes apart, sometimes together. The stories and the venison accumulate. Yesterday I wandered out behind Jed's barn with my rifle and a couple of wrenches to get those wipers off the L-180 and looked up to see a gigantic buck standing against a thin strip of woods two forties away. I dropped the wrench and went for the rifle, but he spun and jumped into the trees. Jed was nearby, so I told him what I had seen and then I stood at one end of the strip while he walked through it end to end, but the buck did not emerge and was probably half a township distant. I returned to the truck. After the wipers, I began removing a chromed strip from the lower lip of the front of the hood. The equivalent strip on my truck was cracked. This one is in perfect shape. It is affixed with four bolts. The nuts on the first three spin right off. The fourth won't budge. I work delicately for a while, then get the bright idea of rotating the strip, rather than the bolt. It is made of cast aluminum and snaps in two immediately.

This is what we call a *"Sunnava!"* moment.

"Walk it off," I can hear Mark say.

○ ○ ○ ○

By the time deer season is over, we have walked miles and miles a day, and sat quietly for hours and hours. We are blessed with this swampy farm of ours, middling for agricultural purposes but perfect for family deer hunting. You can carry me out here blindfolded and give me five steps in any direction and I can tell you where I am, unless you stick me deep in the tamaracks, which are impossible even after thirty years. Life has gotten a little far-flung, but this place and this week have remained the pintle to my wandering.

Anneliese and Amy come up to help with the cutting and packaging of the venison, and it does my heart good to think this could be part of our future. The cutting and storing of the venison, but also maybe a pig, or a beefer. And I fantasize sometimes about looking up from the firewood stack to see Anneliese emerging from a root cellar. It could be looming middle age, but there seem to be more and more of us who feel this way. I remember back in the 1970s during the great Back to the Land rush, seeing the communal farmers make six trips a day past our farm with their little mower and baler, and I remember my father saying, "I wonder how long that will last . . ." and in general it didn't, but today's crop seems by and large more realistic. Anneliese and I have our chicken dreams, but we aren't headed pell-mell off the grid. We have talked to a man named Buffalo about the feasibility of integrating solar and wind power, and so far what I like most about Buffalo beyond his freaked-out Amish beard is that he does not speak of windmills in spiritual terms and he knows what it is to make payroll. We are not under the illusion that we will be farmers of the sort that raised us. In the main, we hope to cultivate a modest crop of self-reliance. Do some things the old way and some things the new. I for one would like to expand beyond freezing to canning. We aren't out to destroy capitalism, just to make it tougher to get our dollar. Take the money saved on car payments and buy an airplane ticket to Panama or bologna for a road trip to North Dakota, or lay off the hydrocarbons and start saving for a recumbent tandem bicycle. Regarding the latter, if we truly wind up together in some legal or other extended fashion, Anneliese has made the recumbent tandem a priority and I have acquiesced, although I have made it clear it will take time because the geek factor there is just astounding.

Once on a bitter cold day when I was still in my teens and not yet past *bagging* things, I dropped my deer hunting hat in the sheep pasture and had to go out after dark to retrieve it. I had killed a doe earlier that day and when I located my cap I switched the flashlight off and tipped my head back to the sky, which was clear of clouds and profuse with stars. Looking at the infinitude and standing in the silent cold that felt as if it was bound to deepen forever, I was swept by the idea of the gutpile somewhere out there freezing in the brush and I had this vague idea that by killing the deer I had cheated the universe of some sort of energy. I walked quickly back toward the house and as always was pleased to see the warm yellow squares of the windows through the pines. But ever since that time I have tried to end at least one day of deer hunting by standing quietly beneath the sky. It is best, of course, when there are stars. You stand there on a frozen patch of the earth twisting away from the sun, and you think of the blood in the snow or on the oak leaves, and in your ears you hear the terminal murmur of your pulse, and you feel very small and uncertain, as you should. In the cold bed of the heavens the stars are saying, *we can wait.*

On a substantially lighter note, Anneliese recently sold the Honda and bought a van for one thousand dollars. It was owned by our friends Mark and Karen. Mark and Karen live simply and travel often. Mark has converted his 1983 Ford Escort to run on cooking grease. Karen was the person who hounded Anneliese to go to the Fall Creek library for a reading by a writer Anneliese had never read.

The van is of the variety they call mini, although I have forbidden the use of this term in my presence, decreeing instead that when I am at the wheel it shall be called the *fambulance*. The engine was recently overhauled and runs like a champ, but the windshield wipers are wired to a standard seventy-nine-cent on/off in-wall light switch, which means you have to get your timing down if you want them to lie flat. You have the option, however, of stopping them at the fullest extent of their outward arc, leaving them parallel to the direction of travel, thus improving handling and aerodynamics.

Chapter 13

DECEMBER

THERE IS A MAN in Eau Claire, Wisconsin, who owns a bar where he will sell you no light beer. It says so in neon right there over the cash register: NO LIGHT BEER. He is uninterested in your opinion on the subject. He has also declared a moratorium on people writing about his bar, and the last time I was in there he told me, "Perry, you're a good kid, but you have to get past Baudelaire."

I didn't have the heart to tell him I'm not even past the Bee Gees.

There is no sign over the entrance of The Joynt, because as a veteran of one corner table once said, this has become the sort of place that doesn't need one. There is an antique barber's chair just inside the door, but you shouldn't just jump right into it, because for years and years Harry sat there, and as the polished granite headstone behind the bar says, HARRY WAS RIGHT. Jack Kerouac once took a leak in the alley out back, and the walls are tiled from ceiling to wainscoting with poster-sized black-and-white framed posters of the jazz and folk artists who have played the narrow tavern (twenty feet wide and three times as deep) in their time—John Lee Hooker, Dizzy Gillespie, Duke Ellington, Koko Taylor, Dave Van Ronk, Stéphane Grappelli, Odetta, Son Seals, Tom Paxton, on and on—but those heydays were over long before I walked into the place for the first time sometime around 1992 and I have no claim on them.

I was led to the The Joynt by a small group of people I met at a local poetry reading. I had just begun attending readings at the invitation of my friend Frank—at that time I knew him only as the editor of a local magazine—and afterward everyone headed for The Joynt. In particular, a corner table just inside the door over which hung portraits of Gary Snyder, Donald Hall, Miller Williams, and John Ciardi. Seamus Heaney once drank whiskey after whiskey at this table, but he didn't read in the bar, so his photo is not hung. There are rules.

For the next five years, I made it to The Joynt one or two Thursday nights a month, as on Thursday night the corner table was occupied by readers and writers. Often they were faculty and students from the local university English department, but you were just as likely to find a carpenter, a single mom, or a local newspaper reporter, and once I listened to a line worker from the plastics factory debate the merits of *Jane Eyre*. The sessions were not without pretension. At times there was enough inflated rhetoric coming off that corner table to resurrect the *Hindenburg*. And we are not talking perpetual lyceum. You had your facile aspersions, injurious gossip, and talk of the Packers. But I would sit there with my Coke (and later, when the corruption had set in, my O'Doul's) and just listen, and the talk would get bluffer and drunker and more ribald, and the air smokier, but there was so much to soak up, so much about books and writers and ideas, that I would return month after month, my rule becoming, the second time Taylor the voluble poet stubs his cigarette on your kneecap, it's time to go home.

There is a temptation with a place like The Joynt to keep recapturing that Thursday night, and when you see the new frat boy whooping and slamming shots in the barber's chair, you want to grab him by the scruff of the neck and say, Son, you are not a drinker, you are a swallower, and furthermore, get out of Harry's chair, turn your cap around, and go sit next to that quiet fellow down there, because he has been here since 10 A.M. for the last ten years and two divorces, and *that,* my young friend, is *drinking,* and good luck catching up. But you don't, because that would be creeping codgerism, and there is grace in letting go, and maybe just maybe Harry's spirit will lead the boy to read a book someday. So after a while I stopped showing up for the Thursday sessions. The Joynt is timeless, it is you who are getting old, and so you ease off and away, knowing you can always come back as long as you show up with reasonable expectations and understand you may have to sit someplace other than the corner. Now I go in there maybe two, three times a year, usually when one or the other of the old crew is coming through town, and it's good to throw peanut shells on the floor, play "Mustang Sally" one more time, and yell old stories over the jukebox. It is good to know there remains a smoky, well-worn place where Nolte the barkeep is capable of making dismissive reference to seventeenth-century French poets while refusing to traffic in panderous beer. And it's just fine to know it's not our place the way it used to be.

○ ○ ○ ○

This year of mine has unraveled in just over a hundred ways, each of them representing a day spent away from home. This schedule is the result of modest accumulation of lucky breaks and the kindness of others, and any attempt to express sufficient gratitude is like chucking quarters into a platitudinous quote generator, housed in the heart though it may be. I say this to establish context for any grumping to come. At every stop and every turn, I have been received with smiles, funny questions, free food, good directions, a hand to help lift another box of books, and a place to lay my balding head. Every time I go to leave, Anneliese and I think of Steve and Sukey, bound to celebrate their sixth anniversary in separate countries, on their second and third tours of duty, respectively, children left at home, no end in sight. It is a time of war, we are told, and although stateside we and our credit cards remain apparently unimpinged, it is damn well a time of war for those two. Sometimes I walk two blocks to the post office and back and upon returning realize I didn't worry about a thing. So when I prattle on here, understand that I have removed my cap.

A year ago I had hit a point of relative clarity. All the usual cranial scrambled eggs remained, but as far as the number of toothbrushes in the bathroom, I was good. I was content as a single man, had rather firmly decided I would limit the scope of my romantic endeavors and most likely never marry, and now here it is December, and held to my desk there beneath the weight of a live 30–06 round is a newspaper clipping with a headline that reads, "Successful Couples Compromise."

I've been going back and forth on marriage, trying to work it out in my head. I'm negotiating two central stumbling blocks: as a wobbly agnostic I have no desire to ask the blessing of some other person's brittle God any more than I feel the need to seek sanction from my own ill-defined list of contenders; nor, as much as I love the Badger State, do I feel any responsibility to declare my love before the state of Wisconsin, although I have downloaded the relevant state statutes and filed them in a manila folder to which I will shortly add the "Successful Couples" clipping. I am not antimarriage, as my life is full of examples of married couples—my parents chief among them—who emanate a depth of calm with each other that is evident of what love can be when it stops acting silly, and indeed I am seeking their counsel whenever possible. I have stayed up late talking to my big friend Billy, who makes his living carving tombstones, and in his fifty years has made the journey from biker bar bouncer to contemplative student of Buddha, and somewhere in between found a wife he treats with affection and dignity. Sometimes I see the old man up the street going for his mail and I remember the day I sat with him and his wife in the Jamboree Days beer tent and he eagerly pulled out his wallet to show me a portrait of them when they met back around World War II—in the photo he is wearing a military uniform—and then I looked at them shoulder to shoulder, both coming in somewhere around a short five feet and grinning with their beers and I wondered what it was to tote that photo around for fifty-plus years and still be so proud to show it with your girl right there beside you. My marriage reservations are not about cynicism or fear. I'm just looking for reasons. There is this anarchist philosopher I have never met but correspond with now and then. When I heard he had gotten married, I sent him an e-mail and asked him why. I guess I believe in it, he wrote back. He said he saw it as a public celebration of a private resolution. He said he meant it when he said *forever*. He said he wanted to make a pledge like that even if in some sense it was ridiculous or impossible.

"Plus," he wrote, "we're both pretty severe advocates of monogamy, you know?"

Anarchists can be such squares.

✿ ✿ ✿ ✿

Today I reviewed my gardening notes, which is a hoot. This year, as with all the other years before, I plotted everything out on graph paper, each little raised bed and patch, not necessarily to scale but done in nice straight lines. The first thing you notice is how many rows of how many plants I have crammed in each bed. Any third-year gardener could have taken one look at the layout and warned me off. There are these earnest little starred notes along the border: *"Thin turnips to 6–8 inches; Parsnips 4 inches."*

Well sure.

There are dates beside each row, noting the day the seeds or sprouts went in, and you can see on some days I went on a binge, planting row after row after row: dill, basil, cilantro, parsley, shallots, lettuce, kale, more basil, cucumbers, parsnips, peas. Another day, and it was just beets and some carrots. There is no equivalent set of dates to note the eventual resolution of each row, although there are some undated comments that provide certain sad clues:

"Thyme—dead-ish."

"Leeks (carryover transplants)," followed by an addition in different ink: *"didn't do well, I think leeks go to seed on yr. 2."*

"Lemon thyme?"

"Shallots." Three years running now, and I have yet to harvest one.

"Pickling cuke." A particularly poignant entry. I made two batches of refrigerator pickles. The first batch sprouted moss. The second batch turned fizzy. I have a Post-it note here from that first batch. It says, *"Mold on the @#! pickles."*

But then there is this entry, faint and barely legible up along one edge of the ill-fated parsnips bed: *"Evening robin rain song."*

Of course I am already thinking about the seed catalog.

My brother John's garden turned out just fine. In particular, his carrots and parsnips did so well he found his root cellar overfull, and I recently came home to find he had left a box of them inside the porch. They were shockingly large. My first impulse was to wave them past a Geiger counter as they had clearly been irrigated with water condensed from the cooling towers at Three Mile Island. I kept turning the parsnips around and around, looking for the label that said Louisville Slugger. Instead of a thank-you note I went to my crisper where the last of my wizened produce lay and, selecting a deeply withered parsnip and a carrot the size of a crayon, tied them in a red satin ribbon and put them on a bed of cotton balls in an empty stationery box, which I then sealed, addressed, and placed in the mail. Within the box I included the following note, printed in frilly Nimbus script:

Thank You
For your recent gift of homegrown
parsnips and carrots.

We were most grateful.

Having said that . . .

We understand the temptation to simply load up on pig dung
and grow the biggest dang vegetables you can. Good for you.

We here at Deliciously Sensitive Farms, however, specialize
in custom-grown, esthetically pleasing vegetable miniatures.
Following in the path of those green-thumbed entrepreneurs
who(m?) sell baby greens to accountants in Manhattan for $57
a pound, we grow amazingly small versions of stuff, year after
year, with depressing consistency. Enclosed in this custom-
packed case please find a selection of our tastefully sized and
undercultivated produce.

Send no money now.

But do please consider the following memberships in our exclu-
sive, earth-centered, utterly Mother-Friendly community-based
sharing circles:

(dues payable in cash or wildly irresponsible credit card)

Stunted Pumpkin of the Month Club
Friends of Wilt
Lentils-R-Us
Sprout-O-Rama
And the
Federation of Anemic Beets

○ ○ ○ ○

I recently stood on a frozen lake inside 3.85 million acres of state-sponsored wilderness with a musher who had been working behind sled dogs for over thirty years but also once spent time in pursuit of a divinity degree, and he said the trouble with America today is not that we're dumb, it's that we're dumb and proud of it. It wasn't that the people I listened to in The Joynt knew it all, or that they had seen it all (although they could account for a pretty wide swath in toto), but that they had remained engaged. It wasn't that they were blue collared or woolly-headed, it was that whether they earned their buck from behind a lectern or on a scaffold, they hadn't allowed their brains to kick into neutral. They hadn't become satisfied. Nobody gave, as they say, two shits if your higher education took place in the ivory tower or atop a pair of drywall stilts, or, for that matter, at the bottom of a glass. This was not on the whole a churchly crew, but they had a fundamental understanding of sin, the greatest of which was to play dumb.

To be sure, there were times you needed tall, tall boots, and every once in a while, if things got too bloated, or someone overreached, Taylor would rare back in his chair and—his one hand triple-tasking a whiskey glass, a cigarette, and a pointer finger simultaneously—would direct our attention across the barroom to the opposite wall and say, "Hey. Hey. Hey." until everyone was looking at the photo-enlarged R. Crumb comic book cover just up the wall from the jukebox and beside the bulletin board with the postcards, bumper stickers, and headline ephemera. On the cover, Mr. Natural—in his robe and wingtips—is riding up the sidewalk on a kick scooter. Flakey Foont and Big Baby are peeping over a wooden fence, and Flakey Foont says, "Mr. Natural! What does it all MEAN??"

And Mr. Natural, just cruising on that scooter, says, "Don't mean sheeit . . ."

We fight and fight to give it all order, have it all make sense, and in the process we tend to sell each other short or cultivate seething caricatures of each other as idiots. All over the country people are painting their neighbors one of two colors. From San Francisco to Manhattan I lost track of how many times some earnest person approached me in a nice quiet bookstore and asked why people in trailer houses were voting, as they say in polite company, "against their economic interests." Having attended a number of caucuses down at the fire hall, I can only conclude the problem is traceable to fifty-seven things, among them a mother of three regularly preyed upon by her line boss and then forced to watch as leading feminists helped zip the pants of a naughty president. When pressed on the issue (and believe me I try to avoid this, as my political views are as delicately constructed as a sugar cube and equally as durable), I said maybe the Donkey's best bet was to have Barbra Streisand send the check and then *put a sock in it*, whereupon they would respond with "Yes, but . . ," which made it clear we should probably talk about cheese curds, an area in which my opinions are informed and durable. Conversely, every time one of my concrete-pouring pals tells me the country is headed to hell in a handbasket thanks to drug-addled media lesbians soft on crime, I am tempted to reply, in my best third-grade-recess voice, "I know you are, but what is Rush?" There are the standard targets, but I swear to you, the scariest thing I heard on my car radio all year aired on a station that had recently switched to a liberal talk format and was running promotional liners, one of which featured a female caller saying, "I used to listen to NPR, but now I listen to Air America in the morning, because they make me laugh." I was pouring hot coffee in my lap at the time, and may have misheard.

But you long for nuance. You long for discussion. You long for the right to muse aloud, then take back what you said and try again. I recently received a very angry note from a man who said he was so incensed by my caricature of a previously mentioned prominent radio talk show host in a book of mine that he had thrown it down immediately and would never read another word I had written, which I take to mean he never got around to the part seven pages later where in a late addendum I retracted the original statement. Bolt from my table, early Sir, and you miss the waffles. I am falling deeply in love with a particular woman because on a regular basis she allows me to say the wrong thing, back up, and try again. She has this *reasonableness*. I love that about her. I love that about anyone. Oh sure, at some point you have to make the call— agreeing to disagree is a privilege predicated on civilized stasis; no less an authority than Louis L'Amour once made the point that straddling the fence will give you a sore crotch; fool me once, all that—but it's nice to hit a few fungoes before the big game without some hothead running in to clobber the guy with the bat.

Dialogue has waxed and waned ever since Socrates took the hemlock, and you know you could always find some grump down at the end of the Athenian coffee bar who would tell you this chowderhead Socrates was no Anaxagoras. And whenever my ears purse at the sound of my own bemoaning, I make it a policy to recall that the halcyon days of yore included the Harding administration. It's just that the two powers currently ascendant are noise and rampant fractionalization. Agents provocateurs aiming to tweak the squares (most of whom are otherwise occupied with paying their rent and fixing your furnace) or marchers *on* the square still draw cameras and microphones, but a sense of diminishing returns prevails. Give your local mechanic a headline and he can tell you how to spin it, left or right. Your hairdresser is a pundit. The plumber arrives with talking points. The yield curve on babble is approaching the perpendicular. Among the piquant conundrums due to face the nation is: How do you whip the folks into meaningful action when everyone is cocooning in like-minded corners of the Web, rather than synthesizing some sort of national unity through the late-lamented mainstream media, which, no matter how you tilt the screen, has had it? The corpse will be wrapped in unsold newsprint, and the viewing will take place online in downloadable form. As a guy who types for a living, I intend to diversify, perhaps into goats, perhaps survivalist chinchillas, but certainly into chickens. I am told there is no longer money to be made in llamas. Additionally, I am pricing solar panels, sawing off my shotgun, and trolling the Y2K sites of yesteryear.

Adjusting on the fly, as one must these days, let me say whoops, *four* powers ascendant. I mean to include transnational corporations and well-armed theocracies, both of which find noise and fractionalization to be reliable sources of fuel and cover. Perhaps the clones of our toenail clippings will view grainy holographs composed of images from Gulf War III: Google Invades. I still sleep at night, because I need the rest.

The culture wars are over and the deconstructionists have won. Which means they've lost. Which—if I grasp the concept—was inevitable. Whatever. The point is, what was once the tool of academic counterculture is now everybody's tool, from your neighbor to The Man. Hector someone to celebrate diversity rather than show tolerance for the noun that it is, and one day he'll call you on the verb and demand you celebrate his. Force the pagans to pledge allegiance to God and they will spank you with a switch cut from your artificial Christmas tree, although the numbers are against them. The nation is daily and at an accelerated rate devouring its own tail. Derrida is dead and we're all deconstructionists now. In further troubling news, Spam is lately available in a single-serving pouch. In the saddlebags of the four horsemen of the apocalypse are sandwiches made of this.

And lest my friend the anarchist grow hopeful, I once responded to an emergency call at an underground music event at which a woman had been stung by a wasp and was going into anaphylactic shock. As we crested the rise in the ambulance, we spotted a barefooted man clad in dust, dreadlocks, and a T-shirt emblazoned with the classic circle-A. "Over here! Over here!" he hollered, urging us to hurry it up with our state-sponsored gear and our syringe full of evil corporate epinephrine. The woman had stopped breathing and when the bagging and adrenaline took effect halfway to the hospital and she roused, I had this teachable moment where I realized the trouble with anarchy is that it is allergic to bees.

○ ○ ○ ○

In my fondest dreams, I sort it all out, using as my chief organizing tool the booklet titled (take a breath), *For Your Convenience in Recording Foods Frozen and Foods Taken Out of Your INTERNATIONAL HARVESTER FREEZER Irma Harding Presents "My Freezer DAY by DAY."* Irma is posted dead center on the cover, looking frank as ever. Gosh, you just know this woman would brook no sass. But once you ate all your peas . . .

The booklet is divided into five categories: Meats, Vegetables, Fruits, Baked Goods, and Other Foods. Each category includes eight columns with the following headers: Food; Number and Size or Weight of Packages (Pints, Quarts, Pounds); Date Stored; Kind of Pack (how prepared); Location in Freezer; and, Removal Record. Instructions are included, and were apparently written by your mother: *Keep this record faithfully . . . it is quickly, easily done . . . the minute or two it will take you to make a record of food frozen and stored, or of food removed from the freezer, will save you many, many minutes and much unnecessary effort later on.*

Really, nothing is left to chance. Consider the instructions for the column labeled Location: *Directions of the compass provide a good way to indicate the portion of your freezer in which any food is stored. Use NE for northeast corner, SW for southwest corner and so on. If you wish to be even more specific, add U for the upper portion of the chest; L for lower portion.*

I am thinking now of the colorful circular plastic sticker that came with my water-filtering pitcher. Solemn as a Boy Scout I put it on my kitchen window over the sink, right where I would see it when I did dishes. Which I still do, and I can report that the first filter cartridge expired September 23, I assume in 1995, since I recall buying the pitcher the year I moved into the house. Later replacement packets included an in-pitcher attachment with an automatic dial that is supposed to track your filter status. I have had mixed results with this latter device, which is to say it is a raging success and the members of the design and engineering departments should get an obscene raise. In specific moments of ground-level clarity, I recognize the dumbing down of the nation for what it is: an act of charity committed specifically for knotheads like me.

So easy, says the booklet, *yet so important.* Irma there on the cover, looking at you with that smile and hairdo somewhere between corporal schoolmarm and encouraging ladypal. And despite my eagerness to please, I just know it would never work out. I have the garden charts to prove it. I'd spend an entire afternoon in the basement, all contents pulled from the freezer and placed in coolers or wrapped in towels and divvied up in laundry baskets, and I'd separate everything out, working within the categories Irma provides, slowly refilling the freezer, noting that the venison roast was at coordinate NWL, the yogurt container of strawberries at SWU, and so on. A little music on the boom box. The day passing overhead. And at some point I would place the last package and now I am thinking what it would be like to go upstairs and sit in my green chair and visualize the freezer one floor down, everything so beautifully squared away.

The following morning after breakfast I would sit at the cleared table with the perfect cup of coffee and consult the guide, tapping my Ticonderoga #2 against my lips as I scan the columns and plan supper. Fish, perhaps. I see I have four packages, half a pound each. Or venison chops. Yes, venison chops. One pound, vacuum-sealed, coordinates NEL. I go straight to them then, but when I tug them from NEL I trigger a landslide in NEU, and during my attempts to rejigger this mess I set off yet another slide in SEU, and by noon I am a wild-eyed frostbitten Sisyphus. Tears of rage in the melted raspberries. *My Freezer DAY by DAY* sodden on the floor, the columns a wretched tangle of cross-outs, Irma's face obscured by a big plop of tomato sauce.

I am told the necessary chips, scanners, and barcodes are already in production so that a dolt like me can monitor the coordinates of my frozen venison sausage from the comfort of my pickup truck. I think of the water filter improvements and never say never. But this misses the point. We keep convincing ourselves we can get a handle on it, despite all evidence to the contrary. In my modest collection of Irma Harding paraphernalia (including a wooden-handled plastic bag sealer that can apparently be used to castrate rodeo bulls) I have two of the day-by-day freezer booklets. Neither has a mark on it. Page after page of barren rectangles. It is better that way. All promise, no disappointment. I find it calming to sit in the green chair and study the unsullied pages, choose one rectangle and use it for a koan. To think of the possibilities. To extrapolate from the freezer to the world, imagine the Removal Record filled in neatly, so when you are tempted to go looking for something that is no longer there, you are spared the disappointment. Do this, and *You'll always know what is in your freezer, and where to lay your hand on it*. We don't ask for much.

It's good to get back to Mark's shop. The furnace is going, all blowy and warm while we work and listen to the radio. Christmas is coming, and this will be our last session of the year. We visited at the kitchen table awhile before coming out. Sidrock crawled around on the linoleum between our legs and ate a Japanese beetle. Mark fished it out with his index finger, lined as always with the residue of his trade. By the time he's toddling full-time, Sidrock will be saturated with a durable set of antibodies.

Mark is attaching the nerf bars today, and they look factory direct: the end of each bar machined to a half sphere, a carefully laid strip of traction tape running the ridge where your foot goes, the frame braces perfectly aligned and angled. I asked him how he did it.

"I just wrote a program in CNC."

"CNC?"

"Computer Numerical Control. I used two-and-a-half-inch round stock and made a full radius rounded end. Just wrote a program in CNC, then put it in the lathe." He pulls a book off the shelf above the workbench and thumbs through it a minute, then places it on the metalworking bench open to Figure 14–8, which is labeled "Turning a Spherical End."

Pointing to the illustration, he says, "The end is proportional to the diameter."

Last night a bunch of the old Joynt crew got together at Taylor's house, specifically *his* kitchen table. Anneliese came along, and it was good to see her welcomed by this bunch that has done so much for me. There was smoking, drinking, and braying, although markedly less than in the old days. The writer John Hildebrand came by late. He took me through *In Cold Blood* the first time I read it a few years back. I recently heard John speak to the fact that a lot of us come from a place where physical labor is the gold standard for judging people, an observation that has helped me readdress a number of my pet obsessions. His *Mapping the Farm* and *A Northern Front* are touchstones of mine, core sources of recalibration. I can also report he is deaf in one ear and, when drunk, develops the malevolent gaze of a quiet man sent to kill you.

If you had gone around that table last night ticking off the Ten Commandments and asking for a show of hands to indicate transgression, you would have thought we were doing the wave. A gloss of the New Testament rules and you would have heard the rotator cuffs snapping. But if you had asked a simpler question—"Who among you is proud of this?"—I think you would have seen no hands, and this is what I love about that crew. Chipped and profane, they have taught me that there is a certain vocabulary you learn only through attrition and heartache. Mercy of the fallen, as the singer Dar Williams put it. The fork in the river where knowledge meets remorse, adds Greg Brown. And in spite of the pomo balderdash all too often generated in academe and held up at arm's length by syndicated foghorners to warn the rest of us off the madness of book learnin', this bunch taught me that calluses and straight-shootin' alone do not a good man make. Here in the welder haze, I'm at a point in my life where I'm trying to figure out where to go and who to trust. Who to surround myself with for the long term or, more to the point, which of them might have me. In these red and blue days, we get gulled into false choices. On a fundamental level, give me a guy who can fabricate nerf bars from scratch, because all the Proust in the world will not clear your sewer line, mangled metaphors nothwithstanding. On the other hand, when some humanities professor knocks around his old lady, you never hear the faculty chair say, Well, yes, but he's a hell of a worker. Or perhaps you do. I have missed some meetings.

I guess, give me articulate sinners who pitch in.

I can't help but notice Mark's sawmill hasn't progressed much since he made room for my truck. His plan was to have it up and running by now, use it to generate extra income on off days and weekends. So far it's been all money in and none out, and he says Kathleen inquires now and then. Although I quite accurately cast him as an ultracompetent superman, he does know what it is to be frustrated. The motorcycle engine didn't work out and he's pulled it in favor of a 60-horse diesel he bought from a guy, but the diesel is currently sitting in a patch of weeds somewhere. And he can't get the hydraulics to work right at all. When he throws the levers, nothing moves the way it should, and the power just peters out. "Some sort of parasitic loss," he says, and I'm thinking, I'd like to hear The Joynt crew parse that one.

While Mark is frogging around under the truck frame, I affix the windshield wiper frames we cobbed from the L-180, and then I hang the spare-tire rack. I grab the bag of carriage bolts and nuts from the plastic tub under the workbench and find the little note I wrote back in April: *"One bolt missing."* The truck was still all rust, Dan had just cut my hair, and it was the day before my first date with Anneliese. Now Anneliese cuts my hair, and I just wave at Dan, but he doesn't mind, I'm going bald and there are still plenty of ladies who need church perms. The truck project has been going slower than we hoped, but life has been roaring right along.

After the rack is reassembled, I mount the front and spare tires, and then mask the sandblasted front wheels for painting. Over the background noise of the furnace and radio, you catch the sound of the ratchet zipping on the backstroke, the staticky splatter of the welder, the tape ripping from the roll. There's not a lot of talking. Before I came over, the fire chief stopped by the house with a package of venison jerky and pepperoni sticks. They're lying there on the workbench, the butcher paper rolled open, and we raid the pile as we work. For dessert, wintergreen lozenges from Farm & Fleet. I don't recall how we got started on the lozenges. They're chemically pink, weirdly addictive, and look like pills for a horse. For some reason, they go good with shop work. We always keep them in stock.

Mark and I get along great, no problems, we just don't have a lot of overlap. We hunt deer together some every year, go ice-fishing some-

times, and five or six times a year we'll end up at the farm for Sunday
night dinner, but beyond that and a few holiday get-togethers, our
contact is pretty limited. When he and my brothers get together, the
talk is all log skidders and compression ratios and welding supplies. I'll
hang around the edges and toss in a joke here and there, and they usually
chuckle, like *Well, you know, it's the best he can do*. Sometimes in the
shop with Mark, I'll get to rambling about the relationship between post-
war industrial design and the evolution of the mustache grille as it applies to the truck-buying habits of Today's Woman, and he has this way of
leaning in with his head cocked a tad, and he holds his eyes a little wide
like you do when you're trying like mad to hold focus, and then I'll notice
his gaze sliding off to the side and then pretty soon he'll just wander
away and start messing around in the parts bin. I figure sometimes after
I leave he walks into the house, slumps in his Mossy Oak recliner, looks
at my sister, and just says, My God.

The reason I like Mark isn't complicated. He is good to my sister, and
he has fundamental talents. If I ran the repair shop we would have odes
on a lugnut, but all your wheels would fall off. That is, if the car would
start in the first place. Mark is, at the end of the day, just a man worried
about paying his bills, raising his kid right, and keeping his wife happy.
He goes to work, fires up his lathe and CNC program, and makes parts
for things we want. Lately here it was luggage racks.

When I put that machinist reference book back on the shelf with the
others, I noticed a fat paperback I hadn't seen before. *The Testament*, by
John Grisham. "That yours?" I asked, as if I had found a fat pink lollipop
in his socket set.

"Yeah. I don't know. I like his books."

"You know, he just started out as a regular lawyer."

"I didn't know that." Which tells you right there he meant what he
said, he likes the books. During a visit to Oxford, Mississippi, I had been
past Grisham's compound, and I started rattling on about the size of the
guardhouse and this thing I had read about Grisham's writing habits and
that I heard he worked out a lot, and you know, there's this little upstairs
bar on the square there in Oxford . . .

Mark cocked his head a tad and widened his eyes.

○ ○ ○ ○

One is hesitant to make declarations unless one is prepared to eat them later baked in pastry, but at the moment the country feels like a chrome-plated diesel wound to the redline, bombing down the road with a big load of nothing. I could be wrong, as the experts say. In the first week of March 2000, after doing my research and thinking it over carefully, I invested my first book advance in tech stocks, and I have been predicting the real estate crash since 1992. Time makes fools of us all, although given enough of it and Einstein's bending light, every Chicken Little gets his shot at Nyah-Nyah. But you have to believe the credit cards are coming home to roost.

In that vein, I should be happy to report, but am not, that after nearly thirty years out of the game, International has resumed manufacturing pickup trucks. Billed as the biggest production pickup in the world, the International CXT weighs 14,500 pounds (you may use the Hummer H2, at 8,600 pounds, for comparison), stands nine feet tall, and is available with a 300-horsepower engine that vaporizes a gallon of diesel every seven to ten miles, all to transport an eight-foot-long cargo bed no bigger than the one on my 1951 L-120. As my brother Jed once said in reference to yet another gigantic dual-wheeled pickup towing two ATVs and a beer-cooler on a lightweight aluminum trailer, "You've got a three-quarter-ton doolie there, son—hook that thing to a *trailer!*" The classic Raymond Loewy logo is absent from the CXT, lost to buyouts and downsizings and focus group specialists who reported that "the brand was also still linked to farmers, which, to many, connoted old-fashioned, unchanging, plain and dull." The logo was updated to have a "bold industrial feel," and strategies were hatched to "get around the farmer image." Mission accomplished: the CXT was written up in *People* magazine, Ashton Kutcher was seen behind the wheel, and it is currently advertised with the slogan, *Trucks that make a bold statement. Namely, that YOU RULE.*

I love trucks. I love big trucks. I love big four-wheel-drive, hammer-down, lay-some-dust-on-the-popples, rifle-racked, ug-ully trucks. Above all, I love *International* trucks. But this CXT is clearly a case of the emperor has no ass.

Seriously. Spam in a pouch.

We pine for dull farmers.

Since I have neither lobbyists nor sufficient mercenaries on retainer to handle the difficulties to come, I will have to satisfy myself with muddling along and engaging in manageable self-improvement projects. Among other acts of citizenship, I must become more stringent about thinning my turnips. The world is interminably nasty and will not be arranged in neat columns. As such, let us keep our powder dry and never again use the phrase, "All is well." Beyond that, living peaceably seems to be the most sustainable form of reprisal.

The year is ending and the truck is still in Mark's shop, the ring is still off my finger, and my garden was the usual mix of compost and plenty, but there is venison and homegrown roast tomato stock in the freezer and a good woman I haven't run off. Whatever the future, I am looking at it from a different perspective. Anneliese has changed my life, for the better certainly, to what extent remains to be seen. I'm still chewing over the marriage issue, recently adding to the manila file a clipping declaring that married people are happier, wealthier, and sexually more satisfied than their unmarried counterparts, although I searched the text in vain for a money-back guarantee. Sometimes I watch Amy playing and think of my jumbled head and my jumbled ways and wonder how the interpolation will proceed, and if I will forget to feed her often. On Christmas Eve we pack the car and drive down to Fall Creek where Anneliese's mother lives on her hilltop farm, and it's a classically cozy evening, sweaters around the crackling fire, the whole bit. After everyone is in bed, I go into the living room to unplug the Christmas tree, and from the window I can see a few dots of light in the valley below, little farmhouses, but off to the northwest the glow of Eau Claire, which is sprawling this way with its bypasses and malls and bulldozing that has been officially and on TV deemed by both the previous and current governors as Good for Wisconsin. That is a debate for another day; more germane to the moment is that one of those governors is a Republican and the other is a Democrat. Sometimes you wonder what all the fighting is about.

Anneliese's mother, Donna, and her husband, Grant, may be moving soon, and there is talk of Anneliese and me taking over the place, should things continue to work out. The idea of leaving New Auburn—sixteen years on the home farm, and now over a decade on Main Street—is more jolting than the idea of getting married, and I am grateful that John Hildebrand recently passed me an essay of his in which he proposes that "sense of place" is not predicated on extended residence but rather is based on our own projections: *this story we tell ourselves about where we belong*. That stood several of my cherished beliefs on their head and snapped their ears but also provided me a way to begin entertaining the possibility of change without unseemly hysterics.

◇ ◇ ◇ ◇

I'm still not crystal-clear on what Nolte was driving at with the Baudelaire thing. To be fair, people had been drinking, and there was a lot of smoke, and we were standing right beside the jukebox, so he had to kind of holler past my right ear, but I do try to take Bill Nolte at his word. I will tell you that by nightfall the next evening I had read some Baudelaire. But I didn't find any clear-cut answer. I recall that at some point during the discussion Nolte said something about ameliorism and the copper roof on Bob Dylan's mansion, so I am going to read up on these ameliorists and see where that takes me, but at some point you have to ditch the interior yip-yap and grab a shovel. Sometimes when I am paddling laps in a demitasse of home-brewed ennui, the concentric circles grow ever-smaller until I spot my own toes, at which point I think of Mr. Natural on his scooter and suspect he may be right. Come to think of it, he may be paraphrasing Baudelaire. But it is time to get past Baudelaire. Nolte will tell you it is good to be confused, rather than soak in your own certitude, be it dark or light. I once asked a veteran Grand Canyon river guide what he had learned after his decades of floating raft loads of tourists down the Colorado. At the time, he was surveying his current sunburned brood as they struggled to set up camp on the beach, and he replied in the instant: "You learn there's a jackass on every trip."

Then he grinned.

"And if you haven't figured out who it is by Day Five . . . it's you."

CHAPTER 14

THE NEW YEAR

I N JANUARY WE went to Mexico. We stayed for five weeks, wandering, hopping buses, staying in cheap hotels, and visiting the family that hosted Anneliese when she was a university student in Cholula. I was dependent on Anneliese to navigate and speak when my seven Spanish words came up short, and it was good for me to need her in that way. I often found myself reduced to asking Amy for words, which she pretended not to know. I scraped together some new vocabulary, but in essence, I was the linguistic equivalent of a big hairy two-year-old.

Much of our trip was spent in the metropolis of Puebla, and inland—Cholula and Taxco—but on our last day, we went to a beach in Zihuatanejo. Amy was a delight, playing in the waves with me, and me sloshing around like a Scandinavian galoot who had been in the ocean only once previously and then for just ten minutes because I was late for an airplane. Anneliese swam with us and then we had what Amy has come to call a "three-hug" in the surf beneath the beating sun, and then we changed out of our wet things and took a taxi to the airport, and later that day, when we unlocked the car in Minnesota, it was 20 below and the foam seats were solid as blocks of cheese.

On the plane ride home, something settled in me. Our seat assignments got split, putting Anneliese and Amy several rows ahead and on the opposite side of the aisle from my seat. Five weeks tramping around

Mexico, swaying through the mountains on fuming buses and cramming into Volkswagen taxis, sharing cheap hotel rooms with nowhere to hide from one another, Amy dragging her pink clearance-rack Barbie backpack every step of the way, and we all still liked one another. During our flight, I couldn't see much of Anneliese, just her blond hair gathered in a twist that revealed the back of her neck as she inclined her head above Amy in the exact motherly cant that has inspired endless versions of *Madonna and Child*—but I studied her from that angle for a long, long while and fixed the image in my head in case I needed it many years from now.

$$\circ \; \circ \; \circ \; \circ$$

When it comes to proposing marriage, I have never understood the concept of "popping" the question. In light of the potential legal and heart-related sequelae, it always struck me that a pre-pop confab might be good. For reasons more civilized than the setting would imply, I once attended a tuba contest during which the proceedings were interrupted by a man who leapt onstage and proposed marriage to a woman who—we have to take his word for it—was his girlfriend, but whom had clearly not been consulted beforehand. In the end, at the urging of the crowd (many clad in lederhosen and hoisting plastic cups of beer), she acquiesced, but her face was a topographic map of nervous blotches and in her haste to exit stage left she scaled the heels of her betrothed. The tubas went straight back to oompah, but it took two verses and a chorus of *Who Stole the Kishka* before I could work the cringe-cramps from my face.

Anneliese and I talked it over first.

At first we thought it would be nice to propose in a natural setting. Beneath, say, a hillside apple tree at her mother's farm, or atop my favorite deer stand. In the end we decided it should be a place neither of us "owned." We figured we'd know it when we saw it.

In March, I was invited to participate in the Tennessee Williams Festival in New Orleans. The three days fit between Anneliese's work schedule, so she flew down with me. On our final night in town, we had dinner with friends, then went for a walk through the French Quarter.

In fact, we strolled. Hand in hand. The town was relatively quiet, and
we were simply meandering when we came upon Preservation Hall.
Having arrived halfway through the penultimate set, we decided to
stay for the final show. At changeover, we packed in with the people
and were lucky enough to find a seat on the second bench back, just
to the right of the center post. Happily, the place was jammed, and
Anneliese had to sit sideways on my lap. The songs rolled out the way
you hoped they would, brassy and swinging, and there was really no
place else in the world, and midway through the clarinet solo of "Yes
Sir, That's My Baby," I looked up to see Anneliese smiling down at me
and, looking back into her eyes with my arms around her waist, I said
into her ear, Would you like to get married? and she leaned to my ear
and said, Yes, would you? and I said, Yes, and then the full band came
swinging back in and there you go.

After we got home from the airport we drove to her mother's farm,
and when we were all three seated in the living room—Anneliese and I
together on the couch, her mother in the big chair—I said, Donna, I am
asking permission to marry your daughter, and Donna said, "My baby!"
and burst into tears.

I took this as perhaps a yes, although when Anneliese moved to the
big chair so her mother might hold her, I have to say I felt like a small
man on a cold island. The room became the provenance of womanly
spirits that spun a swift steel circle around the two of them. I was not
afraid, but I did sit quietly. There were clearly ancient goddesses about.
I hold the scene in my heart as a stern blessing.

It was easier with her father, whom we lured to Anneliese's house
under the false pretense of coffee. When I asked permission for his
daughter's hand, his eyebrows shot up and he said, "Oh!" and then he
said, "Well, I think that's just great."

○ ○ ○ ○

In marrying Anneliese, I will be assuming partial responsibility for a
child. Amy's father and I have had our long talks, and while neither
of us is pretending that this is some sort of neato alternative lifestyle
deal, we have developed what can perhaps surprisingly be called a

heartfelt friendship predicated on the unclouded eyes of a child. We have vowed to do our best as men, and so far, so good. It helps that he has emerged as a thoughtful man of his word who gets Monty Python.

It is my opinion as a long-term bachelor that children are accorded far too much deference these days and that it is time for adults to resume their former responsibilities. That said, rediscovering the world from the perspective of a tyke younger than my newest pair of shoes has chipped away at certain ossifications and let the sun shine in. When two plastic horses get in a fight over cosmetics on either side of a Barbie makeup table, I hear Amy say, in her best horsey voice, "No, that's *my* lip-skit!" When she leans too heavily on one horse and it collapses, she comes up from the floor holding her arm with her face twitching between laughter and tears. Finally, she grins and says, "Now my elbow is flat like a tire!" When time allows, we walk over to the fire hall, where she loves to sit in the trucks and pretend she is one of the "fighter-fighters."

You become privy to the incremental expansion of a brain. I am running an errand with Amy in the car seat behind me when she asks me for a drink. I dig out a bottle of orange juice. Orange juice is her favorite, and at the sight of it, she hugs herself joyfully and says, "Hug, hug, hug!" A few blocks later I check her in the rearview mirror. She is regarding the bottle intently, and then suddenly she looks up, and her face radiates light. "Hey . . . there's *orange* . . . and there's *orange!*" I think it is not insignificant to be present the moment a child discovers that a word—and therefore the world—has more than one meaning.

However sustained the angelic phases, a child is not a Christmas ornament. The capacity for plotting and duplicity seems to come in with that first set of teeth. When my brother Jed remarried, he, too, gained a daughter. Sienna is a few months younger than Amy. The two have become fast friends, usually greeting each other with a full-on sprint to hug. Someday the kinetic energy of their affection will cost one of them a tooth. Recently my car was in the shop, and Jed drove me in to pick it up. We had previously arranged a sleepover for Amy and Sienna, so I put Amy and her Barbie backpack in the car with us, and when Jed dropped me at the repair shop, Amy stayed with him and rode to the farm. She had recently developed the blatant habit of asking for other people's possessions: dolls, toy horses, paring knives, futons . . . whatever struck her fancy. So we'd been working on this, explaining to her that polite guests do not demand gifts. Later, Jed told me that as soon as I exited the car, Amy leaned forward in her safety seat and tapped him on the shoulder.

"I have a secret to tell you."

"Yes?" said Jed.

"When we get to your house."

"Oh-kaaay . . . ," said Jed, not sure what family lid was about to be popped. The rest of the drive passed genially, and by the time they reached the farm, he figured she had forgotten. But as he bent to unsnap her seat belt, she leaned to his ear, and in the gravest of whispers, said, "If there's anything you don't want . . . *I'll take it.*"

I will do my best, which is to say it might be good to start interviewing child therapists immediately. If she trips and falls while chasing her ball, I raise one finger to the sky and declare, "Quoth the Apostle Mark . . ." and she will chime in *"Walk it off!"* Or she will be whining about taking her dish to the sink and I will say, "Zero . . ." and she will say, ". . . *whinage!*" We have also worked up a little act where I load my pocket with invisible dog treats and she goes through this whole production: fetching, rolling over, playing dead. For the big finale I balance an invisible treat on her nose and at the snap of my fingers she snatches it from the air. One day when she comprehends what I have done, I will provide her with the 1-800 number for county social services and take my punishment.

○ ○ ○ ○

The truck is all but ready to hit the road. After switching the system over from six to twelve volts, Mark has run into all kinds of trouble with the peripheral elements of the electrical system. The engine runs fine, but anything attached to a wire is gone nutty. My dad has quite a bit of electrical experience, and he and Mark recently spent an entire day trying to ferret out the gremlins. I was not called in, because I remain firmly in the camp of those who believe that electricity is magic. After a full day's work, they managed to eliminate the majority of problems, but a few remained. Dad called me from the shop.

"The right turn signal works," he said.

"Great!" I said.

"The left one doesn't."

"Oh?"

"It makes the headlights flash."

"Ah."

"But if you switch the headlights off, then it works fine."

I thought about this a moment. "Well, there *is* some good news."

"Oh?"

"I have a left arm. I can hand-signal."

"Yes, but the bad news is, you'll only need the hand signal after dark, when your headlights are on. Whoever's following you won't be able to see your arm."

I can hear Mark in the background. "We could get him one of those reflective bracelets . . ."

I dreaded the idea of picking out rings, and was relieved when Anneliese found her great-grandmother's wedding ring in a drawer and decided to wear that. It is a slim strip of gold with a flat chip of diamond and speaks to me of simplicity and devotion from another age.

My ring—your standard gold band—was given to me by a friend. He wore it for the duration of his relatively brief, or perhaps overlong, first marriage. While standing in the presence of his second wife, I wondered aloud if we were fooling with dire karma here. Think of it as a gift from the kindest, most loyal man I have ever known, she said.

She also said, That first bimbo didn't count.

Glory be, today is the day. The International takes the road. Mark and I are prepared for our first ride, and we are goofy in our glee. We have agreed to keep the first trip simple, and also agree that Farm & Fleet is the only suitable destination. At my request, Mark will take the wheel for the first trip out. This is a matter of honor. If not for him, the truck would be half an inch deeper in my driveway by now, with foxtail sprouting out the wheel wells. Working beside him has been a rare privilege. To watch his hands, to see the way he sizes things up, to see him move forward, make cuts in a blank sheet of steel that would have me dithering for days. More than once I have seen him suss out a problem by looking back and forth from the vintage service manual and the part in question until the image in his head overlaps with the one on the page, and every time I wonder what it feels like to be wired that way. It has also been fun to watch him adapt as a father. I note that up there on the workbench between the Chilton manual and *Motorcycle Basics*, I can now see the spine of the *Giant Coloring and Activity Book,* and over on the floor beside the unfinished sawmill, a multicolored plastic pipe wrench.

We climb in and pull the doors shut. I never tire of the way the latches click, a tight sound that transmits solidly throughout the steel cab. The old truck starts nicely. When it's cold, you have to choke it pretty good, then ease the choke in and the idle out a tad. The Silver Diamond sound is there, sewing machine on the treble end, trucky rumble on the low end, and a touch of midrange putt-putt. Mark guides the gear selector leftward and forward to first gear, and then, pressing the accelerator as he releases the clutch, puts us on our way.

We go grinding up the road. I coach him on how to double clutch, which makes me proud. Like the left-handed threads all those months ago, I am tickled with this rare chance to educate him about something mechanical. Over on the passenger side I mime-shift, working my clutch foot back and forth in the air and pressing the invisible gas pedal flat to the floor. Mark mirrors my act, and with a little grind here and there, we are rumbling on our way. Gosh, it's good to hear the old engine spinning, to see the blacktop slipping under the inverted prow of the hood. The old truck is rolling again, rolling down the road.

Three miles from the house we pull into a corner gas station. Mark pumps the gas while I walk in to pay, and the first person I encounter, this stocky guy wearing an American Breeders Service ball cap, says, "That a 'forty-eight International?"

"'Fifty-one," I say.

"Oh yeah," he said, "my old man had one like that, but I think it was a 'forty-eight."

We pull back onto the short stretch of four lane that runs between Cameron and Rice Lake and Mark works up through the gears until the old six-cylinder is roaring and the cab is thrumming with the vibration of the road. The rubber tires are out of round from sitting so long, and the *whump-whump-whump* feeds up through the suspension. At the first stoplight in town, Mark looks over at me and says, "This is so *cool*."

"Yeah," I say, "and it would be even cooler if we weren't grinning like dorks fresh off the farm."

We pull into Farm & Fleet and park in the middle of the lot. We walk off toward the store but keep stopping to turn back and admire the truck. It's a wonder we don't clothesline ourselves on the cart corral. The yellow parking-lot stripes draw out the green paint beautifully. Parked on the level over the clean asphalt plane, the truck looks trim and lively. All that old steel, looking almost coltish. Cutting back the fenders and pulling the running boards was the right move.

We pick up some odds and ends in Farm & Fleet—including one last bag of wintergreen lozenges—and then it's my turn to drive. It's like easing back into an old familiar chair. It's thrilling the way the muscle memory takes over, how the hand-eye-foot coordination maps are still in the brain, waiting to be accessed like a digital file. My arm weaves the out-of-gear-into-gear movement of the shift lever in between the double-pump motion of my clutch leg. In between gears, I find myself resting my right palm atop the shift knob just the way I used to. Mark and I just keep grinning. It really is getting silly. We take the cure at Twenty-third Avenue, where I have to hang a left across the median and two lanes of oncoming traffic, at which point an air bubble in the brake fluid works itself out of the line and when I push the pedal it slaps straight to the floor like stomping a puffball and we remain

at speed. "Hang on!" I tell Mark, as I pogo my foot up and down on
the brake pedal, trying to drum up the least bit of resistance to our
momentum. "*Pres-stop four-wheel hydraulic brakes . . .*" it says in the
brochure back there on the bench in the shop, "*. . . safe, easy stops
with less pressure required of the driver!*" I am already partially into
the turn and there are cars coming in the opposite lanes, but I figure I
can either beat them or put my faith in our four-point racing harnesses
and jump a small embankment into the grassy ditch of the median. In
the end I calculate our vector on the fly and figure we can thread the
needle. I flatten the accelerator and we careen through the median at
a severe tilt, shooting across both lanes and, seeing no reason to slow, I
keep my foot in it as we straighten and head up the road to home and
at some point I notice the wheels have rounded out and are rolling
smoothly.

We ease up to the shop and head straight for the old truck manual,
opening it at the tab that says *Brake System,* then turn to section A, page 3,
"Bleeding the Lines." Gonna, as Mark says, wanna get right on that.

$$\circ \; \circ \; \circ \; \circ$$

Anneliese and I are attending an abbreviated version of premarital counseling. Frankly, this was not my idea. When Anneliese first broached
the subject, I got quease-inducing images of us sitting criss-cross apple-
sauce and peering at each other's retinas whilst playing patty-cake across
a tub of potpourri. Meanwhile, off in the corner, you'd have some gauzy
guru whacking your chakra. The whole concept clashed with my staunch
sense of by-gum do-it-yerselfness, never mind that after two decades of
captaining my own love boat I was batting a thousand on shipwrecks.
When I heard that Anneliese's mother was behind the counseling idea,
I got the message.

Our adviser is the Reverend Virginia Johnson of the Unitarian Uni-
versalist Church. We are not members of the church, but we have attended a few services, during which Reverend Virginia preached charity
and reason, themes I welcome regardless of denomination. Anneliese
and I desire a thoughtful third party with no partiality for either of us, so
Reverend Virginia seemed a good choice. During our first session, she

had us share the story of how we met. Any happy couple welcomes this opportunity, and I assume a perceptive observer can glean much from how the story is told, from who dominates the narration to who jumps in to correct whom, how often, and with what level of politesse. Meanwhile, the nonverbal reactions accrue.

After a review of our respective family histories we were left alone to complete the Premarital Personal And Relationship Evaluation (a clunky name designed to serve the cutesy acronym PREPARE; the eradication of such forced alphabetical mash-ups is a dream of mine). Because Amy is in the picture, we were assigned the Marriage with Children version. There were 165 multiple-choice questions, each answer represented by a small circle. I selected my circles carefully, scribbling them in thoroughly with a No. 2 pencil, as if my college admission depended on it.

You could tell the interrogation was written up by professionals. Every domestic eventuality was checked off—from kids to dishes, from sex to checkbook. But more to the point, the same question was often posed three or four ways, a sly little trick designed to forestall fudging. Reverend Virginia had told us we were not to speak or consult, and we kept our word, but every now and then one of us would read a question and snort or chuckle, usually because the question covered something we had already hashed out. When the last dot was darkened, we handed in our papers and scooted off to the car and began comparing answers. I suspect the most valuable element of the test may be the conversation that takes place during the ride home.

For our concluding session, we meet Reverend Virginia at a sidewalk café. Over a mocha she reports that our relationship scored as Very Dynamic—which for some inane reason makes me think of the relationship between self-improvement evangelist Tony Robbins and his Super Large Teeth. Reverend Virginia says for all but two categories our scores were synchronous across the board. She gives us the favor of a smile when we correctly predict the two divergent categories: spirituality and—shorthand version here—a couple of loose ends in the family planning department. Biggies. But Reverend Virginia digs to the root of each, pressing us as necessary. In the end she gives us her blessing, comfortable that each understands where the other stands, and that we have established a mutual middle ground. I went into this process suspicious that clinical examination of the fundamentals might take some of the shine off. Instead, I depart feeling more than ever that Anneliese and I are walking shoulder to shoulder. No guarantees, no end in sight, but four good tires and a clear windshield.

I attribute the sweet gravity of the experience to Reverend Virginia's steady hand. We are being advised here not by some cloistral naïf armed with platitudes and well-intended dogma, but rather someone who speaks from a breadth of experience regarding the complications and joy of sworn commitment. For thirty years she and her partner have tended heart and hearth. In sickness and in health, the whole works. They have raised two children, who in turn have given Virginia grandchildren. Virginia has said nothing to us, but I know the rest of her story, which is that this partner of hers is a woman whose health insurance she may not share. This places our coffee klatch in a frame of tin-plated irony indeed.

○ ○ ○ ○

In the essay "Bewildered Snowflakes, We All Are," Annie Dillard writes, *Even lovers, even twins, are strangers who will love and die alone*. I believe this. There is the banal but relevant question of the woman who loves and outlives two husbands. In the beginning, I used Dillard's line against the idea of marriage. Perhaps love is everlasting, but the paperwork expires when you do. Meaning it all comes back to love, and show me the sanction that improves upon the real deal. I remain convinced on this point, but a hundred other little moments—seeing Mark put down his wrenches and go in the house to watch his son, seeing Anneliese and her mother together tearful in that chair—have accumulated and the message seems to be pretty evenly split between *to overthink is not necessarily to be thoughtful* and *for the 365th time this year, son, it ain't always about* you. When I looked at Anneliese on the airplane that day, it began to filter through to me that while busily assembling my crotchety thesis, I have lost sight of the idea that giving your hand to someone

in marriage is above all a privilege. I am sure Reverend Virginia would
agree.

There is also the idea of solemnifying the loyalty of two mortal hearts.
Far from making me sad, Dillard's line does quite the opposite, making
me all the more grateful that Anneliese has agreed to walk beside me of
her own free will, despite the unknown. That when the day comes for
one of us to release the other we will have shared in this life what we
dared hope we might.

◌ ◌ ◌ ◌

Now that it's serious, there have been some changes. Until we know
for sure if we'll be moving to the farm in Fall Creek, Anneliese and
Amy will move in with me, and certain preparations are already being
made. At one point early on in our relationship, Anneliese observed
that my house was less a home than a museum with spiderwebs. She
said this in an amused way, which was nice of her. I have achieved the
decor known as *Midwestern bachelor eclectic,* which is to say the large
finless bass my great-grandfather caught hangs just below the picture
of Johnny Cash, which is nicely accented by the vintage International
pickup postcard. Whenever I am asked why I keep the bass, I point
to the brass plate screwed to the wooden plaque, clearly stating that
Frank J. Smetlak was a *scientific taxidermist.* By this time, however,
they have been drawn around the corner to the crimped and rusty four-
by-eight-foot steel sign hand-lettered with the word TRAILERAMA and
dating from an era when the term spoke not to camp, but campers. The
sign is held securely in place over the stairwell by twenty-three drywall
screws. I rescued it from a ditch during a downpour.

The basement cobwebs were the first to go, and there has been much
scrubbing and clarifying since. "It's not so much a cleanliness thing, but
perhaps a personal hazard," she wrote in one love note, which will give
you an idea of what we both are facing. I have told her she may remove
the billboard panel from the wall in the dining room, but to do so with
consideration, because it took a long time to create a perfectly square
frame from black electrical tape.

The water-filter sticker has been removed from the kitchen window.

Likewise, the framed WD-40 sticker has been removed from the bathroom.

I have slept on a mattress on the floor since 1988 and there is talk of getting a bed.

As a further adjustment, I regularly find myself responsible for feeding a four-year-old when Mommy is teaching night classes. Sometimes Amy gets teary about this, and so we pull out the special candle from Aunt Barbara and light it for Mommy. It helps. Recently I found myself in charge of Amy during a work night at the fire hall. I radioed the chief and said although I was engaged to be married, I could still do whatever I wanted, and tonight I wanted to babysit. He keyed the mic so I could hear all the hooting.

I have also noticed that Anneliese is sneaking boxes of cookbooks into the house. Deeply troubling, as now the unmade recipe count is almost certainly nudging five figures and rising. This is clearly reckless behavior, and was not addressed in the PREPARE sessions. Then again, I have also noticed that she frequently opens a cookbook, says, "This looks good," and makes it. No dithering. I come home to notes that say, *Hay comida en la estufa*, and sure enough it is.

As the wedding draws near, I get a lot of nudges and elbows and raised eyebrows. Especially at the fire department meeting. "Sooo . . . ! How y'holdin' up? Getcha some socks for those cold feet?" That sort of thing. But I couldn't be more relaxed. I am simply easy with the idea that this is right. I don't see any other way to handle it than to go in whole hog, whole heart. Not that you'll find that embroidered on a pillow any time soon.

But I'm good. I'm happy. Ain't skeered.

I mean, I do worry some, sure. I worry because I know the future is ruled by chaos. Never mind the seven-year itch, I've recently had married friends divorce after two decades together. I worry because lately one of my favorite albums—check that, *our* favorite album, the one Anneliese and I most enjoy playing while we are cooking—is Greg Brown's *Covenant*. These are songs of steady love, of enduring love, these are songs about two people providing each other timeworn comfort. These are songs about a man desiring his wife in her *raggedyass old cotton nightgown*. And they are sung by a man headed into his third marriage.

Ach, the future. All I know is what I feel now. I feel like a boy who dreamed he could fly. Then he woke up. And he could still fly.

When my friends Tyler and Jenny got married, they did the ceremony bare bones in a park with two witnesses, and then spent their honeymoon driving around the country visiting the friends they would have invited had they held a standard ceremony. I still remember celebrating their marriage over coffee on my front porch in the morning sun before they drove on, and I'm not sure there's a better way. In this regard, Anneliese and I are going the more standard route. There are relatives coming from as far away as California and Texas, and Anneliese has friends coming from Mexico. My English friend Tim will make a three-day transatlantic flyer just to be here, Bill and Wilda are coming from just outside Nashville, and my friend Gene and his family are driving from Nebraska, reconfiguring their entire summer vacation to include us. We haven't seen some of them for years, and since many of them will be staying through Sunday afternoon, we don't want to blow town Saturday night and lose precious visiting time. So we are planning to stay in the area after the reception and then rejoin some of them on Sunday. We've been trying to decide where to stay that night, and have dithered. We're both cheap, and Anneliese has made it clear we can go the economy route—after all, why splash out for some suite when you're going to arrive tired at 2 A.M., then rise early to catch friends before they hit the road in the morning—but I'm feeling a little pressure here simply as a guy wanting to do right for his girl. Today Anneliese just up and said, "Why not sleep in the back of the International?"

That pretty much salts it.

Three days from the altar, and we have a problem. Because of the indispensable role our friend Minister Katrina played in Anneliese's life long before I made the scene, I happily agreed when Anneliese suggested that Katrina perform our wedding ceremony. Today we have discovered that unbeknownst to Katrina, she has been dropped from the rolls of her church. The story is convoluted and appears to involve venal underhandedness, but in short, her paperwork was pulled over the fact that—I'm paraphrasing Lenny Bruce here—*That Mrs. Johnson, boy, she can throw a baseball just like a man*. T-minus seventy-two hours and we are short one sanctioned officiant. I didn't expect Anneliese to panic, but I was still pleasantly nonplussed when she said, "We'll figure something out." For the umpteenth time, I thought, *Son, you got lucky*. Two minutes later, I was struck with a solution, albeit something other than standard. When I proposed it to Anneliese and she grinned, I thought, *Seriously, son: L-U-C-K-Y*.

Then I telephoned my friend Bob. The same Bob who all those months ago took my call from the parking garage and assured me Anneliese was wonderful and sane.

"Bob. I've got a strange request."

"Delightful!" Bob has a bit of a flair.

I gave him the details. Asked if he could help us out.

"Oh," he said. "I would be honored." I was caught off guard by the softness in his voice, which had shifted from camp to heartfelt in an instant.

○ ○ ○ ○

Two days out now, and this morning I went to see Dan at the Wig-Wam, for one last haircut. It's been a while since I've done more than wave at him. Having Anneliese cut my hair has become one of my favorite little elements of couplehood. I like to sit there quietly while she runs the clippers around my head. The clippers make an annoying clackety-buzz, and I suppose Anneliese uses the time to acquaint herself with newly revealed areas of my scalp, but her tending to me like this seems ceremonial and makes me feel blessed. Perhaps this is why chimpanzees sit still quietly while other chimps pick their nits. But with Anneliese preparing for the wedding, it seemed best to see Dan one more time. We had a nice visit. I updated him on my International and he took me up the street to look at his Scout, repaired and repainted and back on the road. Then I drove to the Minneapolis airport to meet my buddy Tim.

I first met Tim in the village of Great Wyrley, England, on a summer evening in 1984. He didn't say much, as his front teeth had been knocked out in a pub fight the previous evening. We wound up hanging out for hours a day not saying a word, and have been fast friends since, visiting each other seven or eight times over twenty years. The first time he came to America, I was in one of those stages where I had nothing to drive but the International. We hammered all over the place. My brother John took him out shining deer, and I taught him to shoot a rifle. My grandpa took him fishing. I have a snapshot from that visit, he and I leaned against the truck bed, I with a holstered pistol, he with a rifle across his chest. He is wearing a hat John fashioned from a skunk pelt. When we weren't running, we hung out at my apartment, often going for hours without talking. It is an easy friendship.

Five years have passed since we saw each other last, but I spot him at baggage claim immediately. Back in 1984, I was transitioning from feathered look to spritzed mullet, and Tim was stacking his hair high in a modified Thompson Twins mop. Now his hair is going gray and mine is just going. We are both showing some wear. Lines in the face, fewer sharp edges. But our common history keeps us young. He grabs his bag and picks his way through the crowd and says, "O'rright, mate?" as if we were meeting at the pub the same as every Thursday twenty years ago. All this way just to see me married, and in the car I am overfull with gratitude, so full I try to express it. He grimaces and looks out the passenger window, and mumbles, "No worries, mate, no worries." A two-hour drive home and we don't say much more.

Back in New Auburn, he grins to see the International refurbished. We load it with food, thermoses, and bedrolls and drive out into the country, deep into a heavily wooded forty well off the road. Parking at the end of the two-track, we hike deeper in, to this little shack I have. Out here you rarely hear so much as a distant engine. This is my stag party: two old friends, talking some, catching up, but mostly just sitting quiet beneath stars that wrap all the way around the world.

We wake to sun and birdsong coming through the shack screens and walk out to the truck. We have to get started on the trip to the farm in Fall Creek. It's going to take longer today, because we're driving down in the International, and due to the sketchy state of the brake and turn signals, we're sticking to back roads wherever we can. We run a long stretch of County Highway F, all rolling farmland and silos and red barns. Downhill we roar, uphill we chug. Even if we were prone to chatter, the cab noise mostly precludes it. We grin now and then, but mostly we just look out the windows and watch the country go by. I downshift to third to make the last long pull up the hill to the Fall Creek farm, and when we top out and break through the trees the first thing I see is how Anneliese's mother and stepfather have the place mowed and painted and generally straightened. A tremendous amount of toil. I feel full in my chest, and my eyes moisten. I guess there's gonna be a lot of that.

With all the friends and relatives coming in from so far away tonight, we have decided to forego a rehearsal dinner and instead have a potluck and outdoor dance at the farm. We have a rented tent set up over folding tables and chairs under the big old white pine overlooking the valley, and at supper time, people start showing up with a dish to pass. I cannot and do not like to dance, but tonight I am in good hands, because we have hired oddly named band Duck for the Oyster, a four-piece acoustic group that comes complete with Karen the folk dance caller. A pint-sized woman with an encyclopedic repertoire and a willingness to lead, she has us weaving and snaking around the hilltop, one new dance after the other, from old pioneer reels to Guatemalan traditionals. Honestly, anybody can do it, and everybody does, from Amy to Grandma. Only once do I go into the tent to flop in a folding chair for a break. I hold Anneliese on my lap and what we see is our longtime loved ones, our new families, and our good old friends, everybody swinging and bowing and do-si-doing on this hilltop in the late summer evening, the angling golden light perfectly matched to the mellow, woody fiddle. Anneliese stands and draws me by the hand and we get back out there and dance until the dew is down and the sun is gone, leaving the night young enough so that everyone may take to bed and be rested for the morning.

◇ ◇ ◇ ◇

I bought a suit and tie previous to the wedding, which seemed to be not an issue of caving in but a simple matter of respect. Anneliese has mitigated my discomfort by allowing me to accessorize with steel-toed boots—the same pair I wore on our first date. We are to be married at mid-morning. Already Tim and Grant and I have set up the folding chairs and taken them back down because a light rain is falling. Grant had mowed a natural amphitheater halfway down the hillside, but now we're moving operations back up the hill into the tent. I split for the shower.

There is a little room atop the garage overlooking the yard, and it is there I go to dress. John and Jed join me while I'm knotting my tie, and I do my best to avoid saying anything maudlin, as I have given them enough reason for discomfort over the years as it is. We just sit and shoot the breeze and watch the folding chairs filling, slowly at first, then quickly, and then it is time for a grin, a handshake, and a *see-ya-later*. From playing in the dirt to here. Such good men, my brothers. I get a kick out of the idea that they preceded me down the aisle.

And then I am in the quiet house and there is a soft sound in the hall, and Anneliese steps out before me smiling in a grand dress that tumbles all white and brilliant, but it is only her eyes I can see, blue and clear and strong, and then I kiss her—on the lips lightly and quickly, like the little boy who darts in with roses for the beautiful princess and then darts back. Anneliese gives me her hand, the one I have memorized from a thousand miles away, and her three sisters fall in behind us and Amy leads us out the door, across the yard, and within the tent where we stand before a group of people perhaps best summarized as *without whom*.

We implicate them immediately by asking for their communal blessing. Minister Katrina reads a poem by the Native American elder Oriah Mountain Dreamer and preaches a verse from Isaiah. We

give Amy a heart-shaped locket to represent our becoming a family.
Friends and relatives come up in turn to read and sing and speak.
One by one they take their part, and what I keep thinking is you
hope you can live up to the love of people such as these. I know for
a fact there are those in the chairs who wish the service were more
churchly, because they kindly took the trouble to say so, but to them
I can only reply, You, too, brought us to this moment. Besides, I can
see cows from here.

Everything is quite nice—smiles and the occasional dewy moment—
until I stand to thank all three sets of our parents, and am overcome with
weeping. Not the dignified, solitary-tear-down-the-cheek bit, but a full-
on snot-snorking hee-haw. I am grateful that these feelings reside within
me, but for the love of Pete, I wish they'd just sort of ease out now and
then, not slosh over like a kicked bucket. Now I'll never be able to sit
through the wedding video.

Minister Katrina leads us through our vows, and here you are all
alone, looking unblinking into the clear eyes of an open soul, promising
in this profound present to make your heart available forever. We vow
to love and cherish, in good times and in bad, but we also pledge reverence for each other, a word I chose to include specifically based on the
example of my father, who has treated my mother exactly so for forty
years now. We do not promise to obey, as I simply cannot conjure the
circumstance in which Anneliese should be compelled to obey a man
whose proudest achievement since high school is the Unified Laundry
Theory. Furthermore, left in charge of my own destiny I once wiped my
hinder with poison ivy leaves. Reverence, if you really mean it, pretty
much handles *obey*.

My dear friend Gene bears me Anneliese's ring, and Anneliese re-
ceives mine from her sister Marta. Gene is six inches taller than I, and
when he wraps me in a hug I put my head against his chest and I will carry
that moment everlasting. After we exchange rings, two of Anneliese's
friends from Mexico rope us together in a traditional *lazo*. As they drape
it around us, Annie, the farmwife from down the road where Anneliese
used to help make hay, says, "*Mercy!*"

Then it is over. Minister Katrina pronounces us, we kiss, and Amy leads us out of the tent, scattering flower petals from her white wicker basket. She has a lot of petals, and we don't really have anywhere to go, so we take a lap around the galvanized stock tank full of ice and beer.

○ ○ ○ ○

Down at the Masonic Temple the reception kicks off with a family concert. The Temple is one of those classical architectural behemoths run by a dwindling corps of ancient men, but it has a big kitchen in the basement and an intimate auditorium upstairs just perfect for the concert. The stage is draped in a vintage hand-painted backdrop that creates a trompe l'oeil forest.

The concert was a last-minute idea based solely on the availability of the magnificent stage, and we make the best of it. My brother and his barbershop quartet sing their version of "Yes Sir, That's My Baby." I get up there with three musician pals and my guitar with the old-school International emblem on it and perform a handful of songs. Two written for Anneliese, one for Amy, one for my sister-in-law Leanne because I know she puts flowers on the grave of my brother's first wife and she has brought him back to the living, and the last song we sing for the family members who can't be here, including Sukey and Steve. From Anneliese's side of the family we have aunts and uncles and nephews singing in quartets, an aunt performing an early Romantic piece by Adolph Adam accompanied by another aunt on bassoon and a cousin on piano, and, in a performance for the ages, Cousin Paul's bottle band.

When Cousin Paul asked if he could add the bottle band, I figured he'd set up a card table and rap out "Twinkle, Twinkle, Little Star" on six bottles with a spoon and then we'd give him the hook. What we have instead is a windy tour de force featuring somewhere in the neighborhood of twenty-five family members, seventy-nine bottles, one kazoo, costume changes, and a series of sight gags involving the body parts of department-store mannequins. Cousin Paul directs the orchestra in tails using a genuine cork-handled baton. The cumulative sound is a swelling, room-filling combination of calliope meets tuba. Really, you should have heard "Pop Goes the Weasel."

They pulled all this off with something less than two hours of practice, and as I watched Anneliese's ninety-three-year-old grandmother adjust her pearls and hoist her bottle for the chorus of "The Beer-Barrel Polka," three generations of windblown oom-pah behind her, I thought, I hope they can abide me, because I sure like *them*.

All this while, we had friends working downstairs in the kitchen, rafts of people pitching in. We have hired a local woman to cater—beyond her husband, nearly all of her assistants are our friends, drafted to heat things, do dishes, and stock the buffet line with everything Anneliese and her mother test-kitchened with the caterer in advance: Mexican wedding cakes, Mexican meatballs in a chipotle sauce, homemade salsa-fied bean dip, salads, platters of fresh fruit. And outside in the parking lot, the dearest group of roughnecks with which I have ever been associated—the New Auburn Area Fire Department—is set up with their barbecue trailer, charcoaling chicken by the bucketful.

We don't have a head table. I eat my chicken elbow-to-elbow with my old carp-shooting buddy Mills—the funniest man ever to don a pair of plastic hillbilly teeth. He presents me with a carefully wrapped packet of smoked redhorse, and we bore his wife with the same old stories. It's a cozy little lull. Then the band starts and Anneliese and I begin making the rounds, trying to say hello to as many people as possible. I see faces from my childhood, my neighborhood, my high school, my old jobs, my bike-racing days, my community theater days, my nights at the bar days, on and on, and it is a parallel experience for Anneliese. You begin to realize that no thank-you note is going to cut it, and you hope they see it in your eyes. Lieutenant Pam from the fire department, who came down even though she was walking straight-up stiff from back surgery. Uncle Bill, with his table of homemade wine shipped in from Texas. Uncle Stan, who drove his Freightliner all night long to be here. When we came down to eat, the signs of friends giving their time were everywhere—from the arrangement of the tables to the milk cans filled with native flowers and red sumac to the handcrafted envelope box . . . we are blessed, blessed.

○ ○ ○ ○

The reception is at high hubbub when Bob appears at my elbow, raises a conspiratorial eyebrow, and gives a theatrical nod in the direction of the nearest exit. Anneliese locates Marta and Minister Katrina, I round up Gene, and then we all follow Bob upstairs to an anteroom off the hallway, where he reaches within his suitcoat and draws forth a document with such a grand flourish you would expect him to produce the Federalist Papers or a very lacy hanky.

Rather, it is our marriage license.

Some time ago, a young man and woman of Bob's acquaintance met while performing a play. They fell in love and, having succumbed in the footlights, chose to be married onstage at a local theater. The whole production was nearly canceled when it was discovered that the young couple's minister could not perform the ceremony outside of the church proper. Bob, long a fixture of the local community theater scene and ever primed to make a memorable entrance, took it upon himself to go online and do a little research. After obtaining the all-clear from the local clerk of courts and the Wisconsin state attorney general herself, Bob submitted an electronic application to the Web site of Universal Ministries, lately of Milford, Illinois, and within moments found himself selected, appointed, ordained, and granted the power to perform sacerdotal duties including the legal sanction of marriages in the state of Wisconsin. Subsequent to the thespians, he has performed two more. We will be his fourth. Nothing against Universal Ministries, but Bob prefers to announce that he is filling the vacancy left by the late Flip Wilson at The Church of What's Happenin' Now. On a note far less comical, Bob recently collapsed and found himself in the hospital near death, awaiting what is popularly referred to as lifesaving surgery. When Bob's partner of over thirty years showed up, he was forbidden entry until Bob threatened to unplug his machines, leave his bed, and drop in a heap on the sidewalk. Cruel, meet ludicrous.

While everyone celebrates downstairs, you see what has happened here. Anneliese and I are arrived at this moment of pure joy only because three very distinct people—Reverend Virginia, Minister Katrina, and now Bob—shared their time, their hard-won wisdom, and their *love* with us. All that they might guide us to the entrance of a covenant from which they are excluded. Having been given grace, how can I stand for the deprivation of those who freely gave it? This is not about indulging a kink. This is about beating the incalculable odds of finding another human being who loves you and suffers you and holds your most secret secrets in a fortress of commitment, but is shooed from your sickbed. Who knows your shirt size, your white lies, and why you can't sleep some nights, but will receive no privileged consideration under the law without the prodigious tangle of paperwork and high-dollar lawyering required to cobble up an approximation of the real deal. I say let them be joined. If Bob was afflicted with a kink, the other boys would have beat it out of him at recess years ago.

Bob smoothed the license across a lectern, and we each signed off in turn. Bob went last. And then it was real. After a concluding flourish of the pen, he turned, plainly delighted, and pronounced us legally married in the eyes of our assembled friends, the State of Wisconsin, and yes, you bet, The Church of What's Happenin' Now.

We couldn't be happier.

Then it is back to the eating and the dancing and the friends, it goes on and on. At one point before they pack up to go, the entire fire department joins us on the dance floor for a group photograph. I kneel front and center with Anneliese on my knee. The firefighters have all worn their uniform tops, and as we grin I am thinking of the farm in Fall Creek and how we will probably move there, and one day I will go to the monthly meeting to turn in my pager and letter of resignation, and as eager as I am to start my life with Anneliese and chickens, that will not be an easy night. Me and all these people in the blue shirts, we've been through so much together. You could write a book about it.

When we have thanked and paid the dance band (they were a wonderful, old-style big band—I shuffled out a handful of slow dances with Anneliese, and did what I thought was dance when they played, of course, "Yes Sir, That's My Baby"), someone plugs a CD player into the sound system, which starts banging out salsa merengue. Just before the hall rental expires, even this hard-core crew has dwindled, and it all draws down to one table, just a handful of us. Anneliese's mother is laughing and giddy, maybe from the wine, but I think mostly from having it all over with. It is so good to see her happy after all she has done for us. For twenty minutes or so we just sit splayed around a table spearing watermelon chunks from a tray. My tie is undone and Amy is on my lap. Anneliese looks over at us and smiles.

⚙ ⚙ ⚙ ⚙

And then we drive back to the farm where the day began. While Anneliese is in the house putting Amy to bed and changing out of her wedding gown, I prepare the International. It is looking more and more as if we will soon be tending this patch of land. The garden will probably go in over there right about where we stood to be wed. There is a spot just past the garage that seems perfect for chickens, and one corner of the remaining barnyard looks as if it would do for a pig. Or a beef cow. We'll try to keep it realistic. In preparation for the move, we have placed a copy of Gene Logsdon's *All Flesh Is Grass* in the bathroom. We dip into it and dream.

Our plan tonight is to drive out on the ridge running east from the farmhouse and sleep beneath the stars. But there are no stars because the rain is returned and tick-tacking down, so I pull the folded silver Farm & Fleet tarp from behind the seat with the intention of rigging it over the bed. Standing there with a hand on the open door I am thinking: This old truck. Going on twenty years, one of us dragging the other around, and look at us now. Me with a ring on my finger, her painted up and rolling again. She is just sitting there quiet now, the yardlight shadowing dully in her marine green curves. Looking sturdy. Earthy. Handsome for sure. It feels good to think maybe someday soon I'll back her around to the pen and toss some hay to the beefer, or run a load of pig dirt across the yard to the parsnip patch. That maybe one day Amy will learn to double-clutch. The wind is up. I fight with the tarp, try to fix it in place with a set of bungees, but the rain just keeps driving down harder and harder, until I realize there will be no sleeping under the sky tonight.

✿ ✿ ✿ ✿

So I pulled the truck into the big red pole shed and shut the door behind her, and shortly my wife arrived to help me fit the air mattress snug in the truck bed, and then we rolled close in the blankets and fell straight to sleep with the rain roaring on the steel above us.

POSTSCRIPTS

I T TURNS OUT Ron didn't paint the Playboy bunny. I tracked him down with the help of my uncle Mike, who ran a long line of Internationals and used to work with Ron. When I reached him by phone, he was reticent at first, but then he got rolling. He said he bought the truck in Minnesota, where it was sitting in a grove of trees. It had a bad rod, he said, so he took the engine out and overhauled it, which explains why it has always run so well. He says he stripped the engine down to the bare block and found soybean husks inside. He says the yellow flames were on it when he bought it, but that it was a friend who painted it pink, and that the pink wasn't primer, but rather half pints of different paints mixed together. He says his friend was the one who painted the bunny and date on the spare tire rack. He says the tires on the rear end—the ones with the big lugs—were on there when he bought it, and I tell him they're still running. I asked him about the giant gas tank, the one that finally rusted through, and he says it was just a piece of pipe. "I cut some ends and welded'er up," he said.

I moved a lot of stuff with that truck, he said, right before we hung up.

As for Irma Harding, I have learned that she was created by Haddon Sundblom, the commercial artist most famous for creating the rosy-cheeked Coca-Cola Santa, which he would go on to paint for thirty-five years. The model for Irma was a woman named Ann Pfarr. She was paid nine dollars. Over the years, Sundblom also painted a fair share of soda-sipping chickies, and once a *Playboy* cover, further explicating Irma's combined wholesomeness and allure. She was Mrs. Claus with a little of the old *va-voom*.

On November 19, 2005, two years to the month past the projected finish date on the truck project, I walked across the yard in the dark, climbed into the racing seat, turned the key, and punched the starter. She fired right up. Then I drove out into the country, up a dead-end road, and to the far reaches of a plowed field lumpy with snow. Mark met me on the trail, and we walked to deer stands on opposite sides of the swamp and blocked from each other's view by a peninsula of trees. Shortly after daybreak, I caught a movement in the trees behind my stand and when I swiveled to pin it, a big buck startled from the brush in front of the stand. Blowing repeatedly, it ran off before I could raise my rifle, but it was headed in Mark's direction, and soon enough, I heard him shoot. Ten minutes later, I caught another movement, and this time I was able to discern a deer. At the first opportunity, I fired, and took a seven-point buck of my own.

I went to fetch Mark's deer first, bouncing down a winding logging trail, the brush screeching along the fenders and slapping at the windows. I parked at the edge of the swamp and we fetched the deer from the tamaracks. It was a tough drag across the sawgrass hummocks, and we were sweating as we hoisted the buck into the truck bed. Then we motored back over to my side of the woods to load my buck. There was a spot on the trail where the track ran deep into a muddy trough, then climbed steeply. At the apex of the climb, the trail veered sharply left to skirt a tree. Hit the trough too tentatively, and you peter out on the climb, slipping back to stick in the mud. Hit it too hard and you'll over-shoot the hairpin and eat the tree.

I stopped the truck twenty yards shy of the dip, and we sized it up. Mark was in the passenger seat. "Whadd'ya think?" I asked.

"It'll be tight," he said.

"Git'er done," I said.

"Give'er," he said.

Mark grabbed the seat in one hand and the door handle in the other. I double-clutched to second gear and stuffed the foot feed. The engine gathered and the six cylinders went all throaty, and now we were committed, the nose barging forward and down, dead-centered on the bottom of that trough. We hit it with a good head of steam, and just as quickly the wheels struck the upside and the nose bounced skyward. That 127-inch L-120 wheelbase is not so nimble, and the steering wheel was bucking in my hand as the Silver Diamond hood ornament drew a bead on the tree. I kept my foot in it until three feet of bark had disappeared beneath the horizon of the dash, and then I two-fisted the steering wheel, cranking it counterclockwise, hand-over-handing it as fast as I could manage. The ground had frozen overnight, but now we were closing in on mid-morning, and the top half-inch had thawed, coating the solid earth in a mud slick. The rear wheels slipped a quarter turn, the hind end broke loose, and the truck drifted, yawing toward the tree. We were mere inches from broadsiding the trunk. I backed out of the throttle a hair, just enough for the truck to gather and get back on track but not enough to let us sputter and flub the climb. It was a split-second thing, just a centimeter adjustment of my foot on the accelerator pedal amid all the bounce and roar, but it worked a treat, tweaking our trajectory so that we dove past the tree trunk with slim daylight to spare. As we broke clear, I gave a whoop and kept her rolling. Beneath his blaze orange hat, Mark was beaming. We were living the giddy culmination of the truck's resurrection. For every false start, for every change of plan, for everything we never saw coming, for every time we fooled ourselves into thinking we were almost there, here we were, months and miles after rolling the International off the flatbed, roaring for home. Working that truck. *Using* that truck.

Bringing home the *bacon,* no less.

I'm still working on giving Ozzie that ride. We got rained out once and snowed out twice.

$$\circ \; \circ \; \circ \; \circ$$

If you run across an old green International pickup with racing seats and a hefty black brush buster, and maybe a pink Barbie backpack slung from the glove box button, go on around the passenger side and check the spare tire rack. Look in the circle formed by the center of the spare. Maybe you'll see a name painted in there.

If it says IRMA, that's my truck.

About the author

About the book

Read on

Insights,
Interviews
& More ...

Meet Michael Perry

MICHAEL PERRY is a humorist, and author of the bestselling memoir *Population: 485: Meeting Your Neighbors One Siren at a Time* and the essay collection *Off Main Street: Barnstormers, Prophets & Gatemouth's Gator.* Perry has written for *Esquire,* the *New York Times Magazine, Outside, Backpacker, Orion,* and Salon.com, and he is a contributing editor to *Men's Health.* His essays have been heard on NPR's *All Things Considered,* and he has performed and produced two live audience recordings: *I Got It from the Cows* and *Never Stand Behind a Sneezing Cow.* Perry lives in

rural Wisconsin, where he remains active as a volunteer firefighter and emergency medical responder. He can be found online at www. sneezingcow. com.

Raised on a small dairy farm, Perry

About the author

© 2006 by J. Shimon and J. Lindemann

equates his writing career to cleaning calf pens—just keep shoveling, and eventually you've got a pile so big, someone will notice. Perry further prepared for the writing life by reading every Louis L'Amour cowboy book he could get his hands on—most of them twice. He then worked for five summers on a real ranch in Wyoming, a career cut short by his fear of horses and an incident in which he almost avoided a charging bull. According to a series of informal conversations held around the ol' branding fire, Perry still holds the record for being the only working cowboy in all of Wyoming to attend nursing school, from which he graduated in 1987 after giving the commencement address in a hairdo combining mousse spikes on top, a mullet in back, and a mustache up front—otherwise known as the bad hair trifecta. Recently Perry has begun to lose his hair, and although his current classification varies depending on the lighting, he is definitely Bald Man Walking.

Perry has run a forklift, operated a backhoe, driven a truck, worked as a proofreader and physical therapy aide, and distinguished himself as a licensed cycle rider by careening into a concrete bridge completely unassisted. He has worked for a surgeon, answered a suicide hotline, picked rock in the rain with an alcoholic transvestite, been a country music roadie in Switzerland, and worked as a roller-skating Snoopy. He can run a pitchfork, milk a cow in the dark, and say "I don't ▶

66 Perry equates his writing career to cleaning calf pens—just keep shoveling, and eventually you've got a pile so big, someone will notice. 99

understand" in French, Greek, and Norwegian. He has never been bucked off a horse, and contends that falling off doesn't count. He is utterly unable to polka. ∾

Irma & Co.
An Update

MY PINKIE FINGER has gone numb. I am still honing my diagnosis and have yet to pinpoint a cause, but the three leading contenders are a desk, a hill, and a chain saw.

My wife found the desk in a local Goodwill store. It's a solid block of lumber apparently dating from the 1940s. Although I am irrationally attached to my previous desk—an L-shaped particle-board monster swathed in peeling wood-grain contact paper—it is the size of a small aircraft carrier and does not fit the dimensions of my new writing room (we have moved to the farm). I have slouched at the old desk for something like fifteen years. Theory Number One: the numbness may be due to some change of position.

The new writing room is located above a garage (the same perch where my brothers and I watched guests arrive for the wedding). The garage is built into a hillside, and there is a footpath leading down the hill to the house across the yard. Last week I tried to traverse the snowpack in a pair of three-dollar flip-flops and took a cartoon-quality tumble, landing smack on my elbow. As a long-term flatlander used to shuffling about, I need to upgrade my respect for gravity on an incline. The impact was bone-rattling, but I felt no immediate effects. ▶

> 66 Last week I tried to traverse the snowpack in a pair of three-dollar flip-flops and took a cartoon-quality tumble, landing smack on my elbow. 99

5

Irma & Co. *(continued)*

Still, the finger went numb within twenty-four hours. Was there a connection?

The chain saw is an orange Husqvarna belonging to my mother-in-law. She left it behind when she moved away, and I have been using it to make firewood. The Husqvarna has a maddening habit of killing at idle, and I have been unable to tweak the problem, although for the sake of accuracy I must report my only intervention thus far has been to yank the cord in the vicious manner of a man attempting to snap the head off a rattlesnake. The woodlot echoes with inexcusable oaths. The ability to become enraged at inanimate objects is what separates us men from the animals. It's possible I have yanked my pinkie numb.

We made the move to the farm in Fall Creek, choosing the coldest week of the winter to do so. We've been here seven days, and each morning the blue strand of mercury in the thermometer tube is drawn deep below zero. I begin each day with a trip to the woodshed, bringing in enough split oak to keep the fire stoked. At some point later in the day I wander out with the Husqvarna and section up deadfalls. Between cussing fits, I chop what I have cut and stack it for next winter. Based on prior experience, I realize this will likely soon become a chore, but for these first few days it feels hardy and germane. Each morning Amy descends the stairs and makes a beeline for the rug in front of the woodstove. She hugs her knees and soaks up the warmth. In that moment I feel as if I am Providing. Providing health insurance, lunch on time, or useful guidance regarding the boys of tomorrow is another matter entirely. But for now—numb hand or numb brain—I can chop wood.

When the snow clears, I hope to haul wood with Irma. She has been running well, although if I let her sit too long I have to pull the air cleaner and nurse her to life with little dribbles of gasoline down the carburetor, an activity that always makes me feel like one of Louis L'Amour's cowboy heroes tipping a few drops of canteen water down the parched throat of an unlucky compadre left for dead in the desert. Can't give 'em too much too soon, you'll kill 'em, or so the thinking went, and I treat Irma the same.

I took the truck deer hunting again this year, and on regular trips to the village dump. I also made a run out to Jed's farm for a load of pig-seasoned gardening dirt. He filled the bed with one scoop of his skid-steer, and I was amazed at how smooth the old truck rode once the stiff springs flexed beneath the weight of the load.

Irma has suffered one significant cosmetic setback, which could have been prevented had I just painted my own house. The fellow we hired showed up with a motorized scaffold that projected a certain professionalism; sadly, he was a tad fuzzy on how to ▶

66 I took the truck deer hunting again this year, and on regular trips to the village dump. I also made a run out to Jed's farm for a load of pig-seasoned gardening dirt. 99

Before

operate the joystick and in short order managed to slam into the siding, back over a trellis, and finally cap the show by ramming Irma's right front fender, leaving a big smeary dent in Mark's handiwork. This was grumble-making, but as always, one lines it up against the troubles of the world and it goes *poof!*

I haven't brought Irma down to the farm yet. She's a little light on the weather stripping, and I'm waiting for better temperatures. Just now I am recalling that her emergency brake is shot, so I may have to work on that, what with all the hills around this new place. Either that, or keep a piece of firewood in the cab and practice jumping out and slinging the wedge beneath the wheels

After—with a smiling Ozzie ready for his first ride

before the truck shoots off into the duck pond. Me and Irma, getting used to hills.

It's pretty clear to anyone who read *Truck: A Love Story* that there would be no Irma the Truck if there had been no brother-in-law Mark. I have tried to thank him in various ways (Farm & Fleet gift cards, *natch*) but never really got a chance to do it right until the woman who does my taxes got married. Her husband's dowry included a brush-bound but decent short-bed L-model pickup. He offered it to me recently and I was thrilled, but also knew I'd never get it going, so instead Jed and I winched it aboard an equipment trailer and delivered it to Mark as a thank-you ▶

> 66 There would be no Irma the Truck if there had been no brother-in-law Mark. I have tried to thank him in various ways . . . but never really got a chance to do it right until the woman who does my taxes got married. 99

gift for all he did to bring Irma back. Mark says he likes it and can't wait to work on it, but I wonder if maybe I have committed the equivalent of gifting someone with a bucket of uncleaned fish.

In other relevant business, Mark and I convened on a crisp, sunny day late last fall and gave Ozzie that ride. I backed the truck down into a depression to get the rear end closer to the ground, and we bridged the remaining gap with a homemade ramp most certainly not up to code. Ozzie navigated the ramp with a power assist from Mark (the electric motor balked at the incline), and then we locked the wheels and went tooling down the road. I would imagine we violated a number of safe motoring statutes and half the regulations set forth by the Occupational Safety and Health Administration, but there were smiles all around.

More recent, Ozzie completed restoration of his '68 Dodge Charger. He bought the car seventeen years and one life-smashing cervical spine injury ago. Since then both Ozzie and the Charger have been torn down to their very elements and resurrected in altered form. There's a video posted online of Ozzie taking his first ride. I've watched it twice. Right at the end, when the Charger humps up and goes fishtailing down the asphalt, tears fill my eyes. This confirms my emerging

66 Ozzie completed restoration of his '68 Dodge Charger. He bought the car seventeen years and one life-smashing cervical spine injury ago. 99

propensity for weepiness, which I don't mind. It isn't the jet whine of the 426 fuel-injected blown hemi that gets me, it's knowing that Ozzie is cradled in that roll cage, flashing down the road on a journey that had never really ended. Good on ya, mate.

Anneliese and I have been married now for just over two years, and if you ask me, things are going great. Anneliese would agree, although I suspect with certain qualifications of which I remain ignorant unless forced to review them, as in point of fact I sometimes am. She shed tears the day we moved from our house in New Auburn, and I cannot lie, that made me feel better, because I have such affection for the place. If you think the tears may have been less about the romance of place and more a product of endless cardboard boxes, misplaced blender lids, and general duress, I trust you will do me the favor of not coming right out and saying so.

I am currently on my third wedding ring. I lost the first one while delivering a breech lamb (I'm still framing that story and will report back at a later date), misplaced another while making a fire call, and am nowadays wearing one purchased at a head shop in Eau Claire, Wisconsin. It's one of those where if you don't like the shape, you just squeeze it. I like to say the new rule around our house is I am ▶

> 66 I am currently on my third wedding ring. I lost the first one while delivering a breech lamb. 99

Irma & Co. *(continued)*

not allowed to own any wedding ring
that costs more than fourteen dollars.
I buy extras and stash them here and
there.

Having just changed residence, there
is not much more to report. As of yet
we have no chickens. Our first official
farm animal is a guinea pig (currently
nameless) obtained through a hamster
rescue program (yes, America, we do
what we can) and given to Amy as a test
run in responsible pet care as it pertains
to the possibility of one day fulfilling the
dream of owning her own horse. Five
summers spent working as a Wyoming
ranch hand notwithstanding, I am not,
as they say, a *horse guy,* and secretly hope
Amy develops an affection for pet rocks
and existentialism. (I may have been the
only cowboy in all of Wyoming afflicted
with a fear of horses.) For now we are
learning to speak guinea pig, a language
consisting of grunts, whistles, and bubbly
sounds. The key as I understand it is to
differentiate between happy bubbles and
grumpy bubbles. I am glad I do not have
to talk about this with cowboys.

You change your address, you wind
up with a numb pinkie. The butterfly of
chaos does not limit herself to tornadoes.
Half of my ring finger is numb as well,
and having consulted my old nursing
texts regarding dermatomes, I know the
problem is with the ulnar nerve. I'm
going to adjust my office chair, wear
my good winter boots to the office,

take a week off the chainsaw, and see if sensation returns. In the meantime, there are other tasks. We have to order seeds for the garden (production is way up, thanks 100 percent to Anneliese), drag the hog panels from the weeds, fill out mail-forwarding forms, unpack boxes, make all due preparations for the year ahead. Who knows how the world will turn. In April, a baby is due. ∾

> 66 We have to order seeds for the garden (production is way up, thanks 100 percent to Anneliese). 99

An Excerpt from Michael Perry's *Population: 485*

YOU'D JUMP TOO, if you yanked the fire hall door open at three a.m. and found the One-Eyed Beagle in your face. The Beagle is a butcher, and he looks it. His knuckles are knife-scarred. His forearms and biceps are thick with years of hoisting half-beefs from the meat hook to the cutting table. A Fu Manchu mustache brackets his mouth, the gray whiskers dropping straight to his jawline. His lip—even at three A.M., *especially* at three A.M.— is always jammed with chaw. But with the Beagle, it's the eyes that put the startle in you. Several years ago, the Beagle donated a kidney to his niece. He came out of surgery with one eye inexplicably crossed. He jokes about it. His given name is Bob, but he'll be the first to hip you to the One-Eyed Beagle thing.

The eye isn't just crossed, it is out to lunch. If Beagle points his nose south, his right eye shoots due east. Any more easterly, and it is headed back around the bend. And so you understand that when I got to the hall, after running from my house through backyards in the dark, thinking I was the first one there, when Bob rared up right off the end of my snoot, I went a little pop-eyed.

"Scared ya there, Mikey!"

> ❝ The Beagle is a butcher, and he looks it. His knuckles are knife-scarred. His forearms and biceps are thick with years of hoisting half-beefs from the meat hook to the cutting table. ❞

• • •

The Beagle has been on the department
for more than twenty-five years. Only
the chief has more seniority. The Beagle
can tell you some stories. He remembers
back when the old pumper was the new
pumper. He once broke his ankle in the
line of duty. He was running to the hall
in his cowboy boots and landed wrong
when he jumped the ditch. The cowboy
boots are trouble. He'll hit a patch of ice,
and his feet get to *zip-zip-zipping*—it
looks like a hairy version of *Riverdance*.
Lately he has switched to slip-on tennies.
Gives him an edge, he says. The Beagle
and I are pretty much equidistant to the
fire hall door. We're usually the first ones
there, and it's always a little race to see
who gets to drive the van.

Two things you can count on when
you jump in a rig with the Beagle: you're
going to get a whiff of wintergreen, and
you're going to get a detailed report of
exactly where he was and what he was
doing when the page went off. Every
call starts that way. Somewhere in that
first half-mile he'll give you the update.
"Damn, I was just sittin' down to eat, and
I heard Bloomer paged over the scanner
and I said, 'that's our area,' so I started
gettin' my shoes on, and sure enough
here come the page!" "Hell, I was in the
shower, and I thought I heard the damn
pager go off . . ." "Me and the ol' lady . . ."

The whiff of wintergreen you get
because the Beagle doesn't make a move
without a fresh plug of Kodiak. After all
these years, the cud slips in between his
cheek and gum like an afterthought and
stays there. You think he'd take it out at ▶

mealtime, but he says no. "I could eat an apple, I got that big a pouch in there," he says.

You'll smell the wintergreen, but you'll never see the Beagle spit. He is of the man-enough-to-chew, man-enough-to-swallow school. My pancreas cramps up at the thought of it. He says his dad called him the other day, told him to bring over a tin of Kodiak. The Beagle found this strange, because his dad is always getting on him to quit the chew. Turns out dad's dog had worms, and he wanted the tobacco for a de-wormer. The Beagle said it was a tussle, but they got a wad of snoose down Shep's gullet.

"That dog shit worms for a week!" said the Beagle. We were sitting around the fire hall at four in the morning, back from some call.

"And what does that tell you, Beagle?" I had to say it, to set him up.

"I sure as hell ain't got worms!" ～

Have You Read?
More by Michael Perry

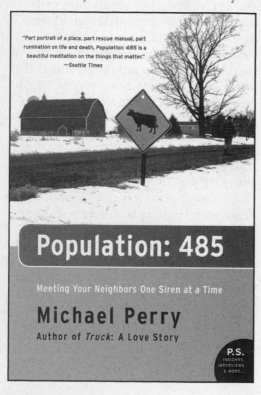

POPULATION: 485

Here the local vigilante is a farmer's wife armed with a pistol and a Bible, the most senior member of the volunteer fire department is a cross-eyed butcher with one kidney and two ex-wives (both of whom work at the only gas station in town), and the back roads are haunted by the ghosts of children and farmers. Michael Perry loves this place. He grew up here, and now—after a decade away— he has returned.

Have You Read? *(continued)*

Unable to polka or repair his own pickup, his farm-boy hands gone soft after years of writing, Mike figures the best way to regain his credibility is to join the volunteer fire department. Against a backdrop of fires and tangled wrecks, bar fights and smelt feeds, he tells a frequently comic tale leavened with moments of heartbreaking delicacy and searing tragedy.

"Swells with unadorned heroism. He's the real thing."   —*USA Today*

"Part portrait of a place, part rescue manual, and part rumination of life and death, *Population: 485* is a beautiful meditation on the things that matter."
                    —*Seattle Times*

OFF MAIN STREET: BARNSTORMERS, PROPHETS & GATEMOUTH'S GATOR (Essays)

Whether he's fighting fires, passing a kidney stone, hammering down I-80 in an eighteen-wheeler, or meditating on the relationship between cowboys and God, Michael Perry draws on his rural roots and footloose past to write from a perspective that merges the local with the global.

Prior to writing the beloved memoir *Population: 485*, freelance journalist Michael Perry wrote essays on such diverse topics as big-rig truck driving, country music, butchery, farming, nursing, and the many facets of small-town America. *Off Main Street: Barnstormers, Prophets & Gatemouth's Gator* is a collection of Perry's offbeat reporting, and is confirmation of his one-of-a-kind worldview and deeply personal and humane insights.

In "Branding God," we witness Perry working on a cattle ranch and recalling time spent under the spell of a fire-and-brimstone preacher by the name of Brother Timothy. In "Rolling Thunder," Perry rides with a convoy of Vietnam veterans on their chopper-rigged march to the nation's capital. In such pieces as "A Way with Wings," "Swelter," and "Manure Is Elemental," Perry reflects on his own boyhood spent in rural Wisconsin amid everyday characters of the American Midwest.

Ranging across subjects as diverse as lot lizards, Klan wizards, and small-town funerals, Perry's writing in this wise and witty collection of essays balances earthiness with poetry, kinetics with contemplation, and is regularly salted with his unique brand of humor.

"Michael Perry is like a sensitive, new-age Hemingway." —Salon.com

Don't miss the next book by your favorite author. Sign up now for AuthorTracker by visiting www.AuthorTracker.com.